Community Quality-of-Life and Well-Being

The Community Quality of Life and Well-being book series is a collection of volumes related to community level research, providing community planners and quality of life researchers involved in community and regional well-being innovative research and application. Formerly entitled, Community Quality of Life Indicators: Best Practices, the series reflects a broad scope of well-being. Next to best practices of community quality-of-life indicators projects the series welcomes a variety of research and practice topics as related to overall community well-being and quality of life dimensions, whether relating to policy, application, research, and/or practice. Research on issues such as societal happiness, quality of life domains in the policy construct, measuring and gauging progress, dimensions of planning and community development, and related topics are anticipated. This series is published by Springer in partnership with the International Society for Quality-of-Life Studies, a global society with the purpose of promoting and encouraging research and collaboration in quality of life and well-being theory and applications.

More information about this series at https://link.springer.com/bookseries/13761

Scott Cloutier · Sara El-Sayed · Allison Ross ·
Melanie Weaver
Editors

Linking Sustainability and Happiness

Theoretical and Applied Perspectives

Springer

Editors
Scott Cloutier
School of Sustainability
Arizona State University
Tempe, AZ, USA

Sara El-Sayed
School of Sustainability
Arizona State University
Tempe, AZ, USA

Allison Ross
College of Health Solutions
Arizona State University
Phoenix, AZ, USA

Melanie Weaver
Phoenix, AZ, USA

ISSN 2520-1093 ISSN 2520-1107 (electronic)
Community Quality-of-Life and Well-Being
ISBN 978-3-030-89558-7 ISBN 978-3-030-89559-4 (eBook)
https://doi.org/10.1007/978-3-030-89559-4

This Springer imprint is published by the registered company Springer Nature Switzerland AG
The registered company address is: Gewerbestrasse 11, 6330 Cham, Switzerland

Contents

Chapter 1
Introduction to Linking Sustainability and Happiness: Theoretical and Applied Perspectives

Scott Cloutier, Sara El-Sayed, Allison Ross, and Melanie Weaver

Welcome

Hello Friend!

Thank you for taking the time to pick up or link to this book. When we set out on this journey, the intention was to integrate scientific evidence, spiritual aspects, traditional knowledge, mysticism and other means and methods of exploring the links between sustainability and happiness. What has been delivered is an exciting array of studies from an international group of scholars that support prior findings, open new inquiries, acknowledge limitations, and suggest a bright future for sustainability and happiness research. What we can offer you is a toolkit for using the book to inform your own interests in sustainability and happiness research. First, we ask you to relax your mind to make space for questions that emerge from other ways of thinking, knowing, and being. Some of the ideas posed within the book are likely familiar to you, and others are very likely not—notice the spaces of resistance, inquiry, and knowing you hold, as they are valuable teachers. Second, we ask you to open your heart to the possibilities that relating sustainability and happiness can produce. There is a deeply knowing place that rests within the heart space as an important contributor to this conversation and exploration within the pages of the volume. Finally, we request your humility and appreciation as we thank our contributing scholars for

S. Cloutier (✉) · S. El-Sayed · A. Ross · M. Weaver
Arizona State University, Tempe, AZ, USA
e-mail: Scott.Cloutier@asu.edu

S. El-Sayed
e-mail: Sara.elsayed@asu.edu

A. Ross
e-mail: Allison.Poulos@asu.edu

M. Weaver
e-mail: Melanie.Weaver@asu.edu

S. Cloutier et al. (eds.), *Linking Sustainability and Happiness*,
Community Quality-of-Life and Well-Being,
https://doi.org/10.1007/978-3-030-89559-4_1

putting together a shared space of open inquiry and exploration of a relationship that is inherently important to humanity's future with our teacher and guide, Mother Earth.

With Love,
Scott, Sara, Ali, and Melanie

1.1 An Honoring

Before we delve into the introduction of the book, we want to honor all the beings before us that have contributed to and explored this space. Our work is in no means beyond these works, even with seemingly new findings. We see the book as a contribution to our responsibility to be aligned with our planet and an opportunity to (re)align with our ancestors to find a shared path toward sustainability and happiness.

Below, we will contextualize our terms of interest, as well as some of their purported linkages, as the terms are more deeply defined in the included chapters. The sections within this introduction are not exhaustive, in any way, and are more fodder for the future of what possibilities may come when bringing sustainability and happiness research together. We seek to push the edge of what is commonly discussed and cited, opening the field to new ways of knowledge (re)generation and being.

1.2 What is Sustainability?

Sustainability is a word that generates more questions than answers, as it should be. While many individuals, communities, organizations, and institutions acknowledge we are pushing the limits of our shared planet, no universal definition of sustainability or means of putting it into action has been agreed upon. Perhaps one of the most often cited is that from the Brundtland Report (World Commission on Environment and Development, 1987), defining sustainable development as a general responsibility for our future generations' needs, while thinking more deeply about our own. The definition is useful in a sense, yet, it is limited by its lack of acknowledgment of the many prior cultures, places and their nuances (Schilling, 2018) from which sustainability is derived—as are many papers citing the definition. Specifically, the Brundtland Report definition is built upon the work of cultures that were better attuned to natural systems, reciprocal living with the systems that support human life, and humanity's responsibility to a planet shared with all our kin (LaDuke, 2005) (e.g., the beaver people, the hummingbird people, the Wind, and Father Sun). Interestingly, the Brundtland definition was borne from a group of people struggling to live sustainably and on the path, perhaps, to becoming more disconnected from Mother Earth than anytime in humanity's short window of existence. And, while not romanticizing past cultures who certainly worked with overconsumption and unsustainability, there are great offerings from lessons they learned. For instance, the Iroquois Nation and its Great Law of Peace (Williams, 2018) is often credited for thinking about Seven Generations ahead when making current decisions that require resources. Other indigenous teachers have emerged reminding us of our responsibility to Mother Earth as stewards, caretakers, and beings observing and adhering to the law of nature (Lyons, 1991, 2016). Such people and nations serve as our elders, our teachers, and our leaders from which we can derive wisdom and knowledge to move toward a sustainable future.

In general, we suggest that sustainability can be thought of as both a noun and a verb. The former begets a destination where humanity lives in a reciprocal relationship with Mother Earth, honoring and acknowledging our reliance on her. It is a place where a concept like regenerative development, or learning from and attuning to land's natural rhythms to cultivate abundance, is the norm. More, it is a place where traditional knowledge is (re)learned and passed onto our children and children's children in written word, yes, but also in song, prayer, practice, and many forms of creative expression.

Sustainability as a verb is a process of exploration, growth and renegotiations. It touches on a spiritual (re)connection with the laws of nature and rhythms of Mother Earth. There is a myriad of pathways to finding this (re)connection, coming from any combination of intellect, heart, and spirit. In itself, sustainability verb is the process of reaching sustainability noun. What is important, is not how the verb is actualized—it is more important toward what the process is moving. The intellect is invaluable here, as it can generate great interactions and interventions that honor sustainability noun. We suggest that our work, as academic scholar practitioners, is to create the spaces where noun can be (re)visioned and verb can be (re)aligned with moving toward that space. Our book is an offering toward such an approach.

1.3 What is Happiness?

Ah, the age-old question that has been explored for thousands of years. A seemingly endless pool of happiness research and definitions exist but, for the purposes of brevity, we will leave this to the chapters within. Generally, we, the editors, refer to happiness as a sense of wellbeing grounded in a life of joy and contentment. Much like sustainability, the term can be thought of as both a noun and a verb. Happiness noun is also a destination, and we suggest that this space might not be far off from the sustainability noun space. In other words, we suggest happiness noun is a place that focuses on regeneration from connection with natural systems and a reciprocal relationship with Mother Earth and all her beings. Happiness verb, on the other hand, is the process of moving toward noun. We clarify this relationship with what has been referred to as two distinct forms of happiness (Deci & Ryan, 2008): hedonia and eudaimonia.

Hedonia, simply put, is pleasure. Cloutier (editor 1) has long purported that hedonia is a significant contributor to our shared lack of sustainability, in the sense that dominant society has created an array of pleasurable experiences—and we are all complicit. In short, the archetypal modern human can find numerous forms of pleasure close at hand. Such a design has many implications, the most significant of which is overconsumption feeding into climate change. More, the happiness derived from a life of short-term dopamine spikes can be an arduous path from which long-term happiness is not found. Both our biological machinery and spiritual hunger reinforce this idea. The former is connected to biological concepts of habituation and adaptation—as we obtain the same form of pleasure, our reinforcing physiological signals become weaker, encouraging the seeking of more and new forms of pleasure. The idea has been referred to as the hedonic treadmill (Lyubomirsky, 2011). The latter, spiritual hunger, is what underlies our desires to find pleasure in the forms of soft (i.e. technology, video games, overeating) and hard (i.e. drugs, alcohol) addictions. In essence, we seek deeper human experiences like community, connection, and feeling loved, and temper these deeper needs with short term pleasure gains. Such experiences touch on the edge of eudaimonia or, simply, happiness derived as meaning and purpose. Eudaimonia is a happiness that, when fully activated (through meaningful relationships, time in nature, time volunteering, and being of service etc.), may reduce our dependence on the planet through longer lasting experiences of happiness and the reduced need to consume and put excessive demands on our planet.

1.4 Sustainability and Happiness Linked

A limited number of studies have directly linked sustainability and happiness (Cloutier et al., 2020; Cloutier et al., 2017; Cloutier & Pfeiffer, 2015; O'Brien, 2013; O'Brien, 2008; Zidansek, 2007) and the intention of this book is to broaden the number of studies that do. It is also the intention to bring together a collaborative working group of academic scholar practitioners moving sustainability and happiness research forward. Simply, for the purposes of our introduction, we suggest that sustainability noun, a destination ripe with regeneration, and happiness noun, a destination ripe with nourishing experiences, are not far off from one another. Consequently, happiness and sustainability verbs can be thought of as complementary processes to actualize a sustainable future. We will not go deeply into these relationships, as our contributors have brilliant perspectives of their own to share. A quick summary of our contributed chapters is shared next.

1.4.1 Volume Contributions

As the title of the book states, we sought contributions linking happiness and sustainability from theoretical and/or applied perspectives. The idea behind such a divide was two-fold: (1) to provide space for refined and new theories that expand the way happiness and sustainability are related, and (2) to open space for applied research and practical perspectives. The volume has been divided as such, with the first section comprising four chapters of theoretical perspectives, and the second section comprising six chapters of applied perspectives. In total, our book and its contributions represent a diversity of countries, cultures, and approaches. Many "traditional" perspectives are covered in our book and new ones are introduced, including double dividend, regeneration, community productivity, buddhist sustainable development, and more. The chapters are briefly summarized below.

Theoretical perspectives:

(1) *Achieving Community Happiness and Well-being through Community Productivity* by Maria Spiliotopoulou and Mark Roseland: makes a convincing case that community happiness can be achieved through a holistic perspective of enhanced community productivity and quality of life.
(2) *Sustainability and Happiness as "Promoting Life"* by Jacob Bethem: suggests that "promoting life" supports both happiness and sustainability, while exploring what the value of life is.
(3) *Buddhist Sustainable Development: Inner Happiness as a Direction for Sustainable Development* by Sauwalak Kittiprapas: posits Buddhist Sustainable Development (BSD), where inner happiness adjustments, rather than external interventions, are central to achieving sustainable and happy individuals and communities.
(4) *Mahatma Gandhi's Sarvodaya (Welfare for All) as a Way to End Violence and Achieve Happy, Sustainable Societies* by Jorge Guardiola: details *Sarvodaya* as a means of moving toward a sustainable and happy future for all beings.

Applied perspectives:

(1) *Where is the double dividend? The Relationship Between Different Types of Pro-environmental Behavior and Different Conceptions of Subjective Well-being* by Nazaret Ibáñez-Rueda and Jorge Guardiola Wanden-Berghe: assesses the relationship between pro-environmental behaviors and subjective well-being amongst university students.
(2) *Are the Sustainable Development Goals the Compass for a Happier Society?* by Leire Iriarte from El Buen Vivir: explores how the SDGs and various environmental indicators interact with the happiness and well-being measures at the country level and reflects on some of the implications.
(3) *Are Turkish Housewives Happy? A Qualitative Approach* by Shoirakhon Nurdinova: identifies interesting factors affecting women's decisions to participate in the labor market, or not.
(4) *Inner Happiness and Environmental Preservation in a Green-Space Community* by Sauwalak Kittiprapas: investigates happiness and sustainable development (environmental preservation) relationships in Thailand.
(5) *How Sustainable is Mobility in Cities Branded the Happiest?* by Alshimaa Aboelmakarem Farag: explores the relationships between happiness indices and sustainable mobility measures in the world's happiest cities.

(6) *Linking Subjective Wellbeing and Pro-Environmental Behaviour: A Multi-dimensional Approach* by Christian Krekel and Alberto Prati: explores the relationship between happiness and pro environmental behavior.

You will hear more from us within the concluding chapter. We hope you enjoy the book and the many perspectives within, and glean as much information, knowledge, and wisdom as we did while putting it together.

With Love,

The Editors.

References

Borgonovi, F. (2008). Doing well by doing good. The relationship between formal volunteering and self-reported health and happiness. *Social science & medicine, 66*(11), 2321–2334.
Cloutier, S., Angilletta, M., Mathias, J. D., & Onat, N. C. (2020). Informing the sustainable pursuit of happiness. *Sustainability, 12*(22), 9491.
Cloutier, S., Karner, A., Breetz, H. L., Toufani, P., Onat, N., Patel, S., & Carlson, C. (2017). Measures of a sustainable commute as a predictor of happiness. *Sustainability, 9*(7), 1214.
Cloutier, S., & Pfeiffer, D. (2015). Sustainability through happiness: A framework for sustainable development. *Sustainable Development, 23*(5), 317–327.
Deci, E. L., & Ryan, R. M. (2008). Hedonia, eudaimonia, and well-being: An Introduction. *Journal of Happiness Studies, 9*, 1–11.
Fowers, B. J. (2016). Aristotle on eudaimonia: On the virtue of returning to the source. In *Handbook of eudaimonic well-being* (pp. 67–83). Springer.
Holder, M. D., Coleman, B., & Wallace, J. M. (2010). Spirituality, religiousness, and happiness in children aged 8–12 years. *Journal of Happiness Studies, 11*(2), 131–150.
Howell, A. J., Dopko, R. L., Passmore, H. A., & Buro, K. (2011). Nature connectedness: Associations with well-being and mindfulness. *Personality and Individual Differences, 51*(2), 166–171.
LaDuke, W. (2005). *Recovering the sacred: The power of naming and claiming*. South End Press.
Lyons, O. (1991). An Iroquois perspective. In B. Willers (Ed.), *Learning to listen to the land* (pp. 202–205). Island Press.
Lyons, O. (2016). Listening to natural law. In L. Vaughn-Lee (Ed.), *Spiritual ecology: The cry of the earth* (2nd ed., pp. 13–20).
Lyubomirsky, S. (2011). *Hedonic adaptation to positive and negative experiences*. Oxford University Press.
O'Brien, C. (2008). Sustainable happiness: How happiness studies can contribute to a more sustainable future. *Canadian Psychology/psychologie Canadienne, 49*(4), 289.
O'Brien, C. (2013). Happiness and sustainability together at last! Sustainable Happiness. *Canadian Journal of Education, 36*(4), 228–256.
Ryan, R. M., & Deci, E. L. (2001). On happiness and human potentials: A review of research on hedonic and eudaimonic well-being. *Annual Review of Psychology, 52*(1), 141–166.
Schilling, J., Froese, R., & Naujoks, J. (2018). "Just Women" is not enough: Towards a gender-relational approach to water and peacebuilding. *Water Security Across the Gender Divide*, 173–196.
The World Commission on Environment and Development (The Brundtland Commission). (1987). *Our common future*. Oxford University Press.
Waterman, A. S. (2007). On the importance of distinguishing hedonia and eudaimonia when contemplating the hedonic treadmill. *American Psychologist, 62*(6), 612–613.
Williams, K. P. (2018). *Kayanerenkó: wa: the Great Law of peace*. University of Manitoba Press.
Zidansek, A. (2007). Sustainable development and happiness in nations. *Energy, 32*(6), 891–897.

Chapter 2
Achieving Community Happiness and Well-Being Through Community Productivity

Maria Spiliotopoulou and Mark Roseland

Abstract Since the middle of the twentieth century, social isolation and unhappiness rates have been increasing almost in parallel with an increase in house sizes and material possessions. Is there a way to develop a happy community with a high quality of life and well-being? We propose that it is possible to tackle the realities of individual and community isolation through whole systems thinking coupled with improvements in the quality of life, with less stuff, enhanced social connections, and ideally reduced and more efficient resource use. In this chapter, we discuss the need to transition from current sustainable community development theory and practice towards a productive and regenerative community. Following a brief review of the emerging concept of community productivity, we argue that achieving community happiness and well-being is possible through increased productivity of multiple, tangible and intangible, community assets.

Keywords Sustainable communities · Community productivity · Community development · Community well-being

2.1 Introduction

A happiness-related paradox is often cited in the literature: rich, developed-world communities facing increasing rates of depression and life dissatisfaction. Although referring to a controversial and multi-faceted debate, a diminishing returns relationship between happiness and income can be found in today's highly urbanized society (Frey, 2018). To paraphrase *The Doors*' song *People are Strange*, people seem strange, ugly, and wicked when we feel strange, alone or unwanted. More and more urban dwellers report feeling like this every day, regardless of their income or house

M. Spiliotopoulou (✉)
Simon Fraser University, Burnaby, Canada
e-mail: mariaspi@sfu.ca

M. Roseland
Arizona State University, Tempe, USA
e-mail: Mark.Roseland@asu.edu

S. Cloutier et al. (eds.), *Linking Sustainability and Happiness*, Community Quality-of-Life and Well-Being,
https://doi.org/10.1007/978-3-030-89559-4_2

size. Since the middle of the twentieth century, social isolation and unhappiness rates in western societies have been increasing almost in parallel with the increase in house sizes and material possessions. Is there a way to develop a happy community with high quality of life and well-being? We propose that it is possible to tackle such issues through whole systems thinking coupled with improvements in the quality of life, with less attachment to material products, enhanced social connections, and ideally reduced and more efficient resource use. The links between such enhancements in quality of life, well-being, and the concept of community productivity—which advocates for systems thinking in community policy and development—will be elaborated on below.

In section two we provide information on the methods used for this chapter and in section three, we briefly explain why a transition from current sustainable community development theory and practice is necessary. In section four, we present the emerging concept of community productivity and we provide examples of relevant metrics and current initiatives from around the world. Finally, in section five, we argue that achieving community happiness and well-being is possible through increased multi-factor productivity.

2.2 Methods

To explore the issues associated with sustainable community development and the transition to productive community development, we first studied academic articles through a traditional, theoretical literature review process. Starting with seminal academic work on sustainable community development and urban sustainability, we expanded the search through Simon Fraser University Library's search tool[1] using search terms such as *urban sustainability debates*, *local sustainability case studies*, *urban systems theory*, *urban sustainability frameworks*, *urban metabolism*, *urban governance*, *urban resilience*, *urban resource productivity*, *social productivity*, *ecological productivity*, *urban regeneration*, *regenerative sustainability*, and *sharing economy*.

In an iterative way and concurrently with the scholarly literature review, we also consulted non-academic sources such as scientific reports, non-governmental and international organization documents, handbooks, websites, and edited books. All references were evaluated for credibility and soundness, in terms of author, publication venue, content, and methodology. This chapter provides a succinct review of conceptual influences and limitations in sustainability and in productivity theory and practice, and demonstrates interdisciplinary influences as they pertain to the goal of increasing community happiness and well-being.

[1] The current web link for the SFU library's search tool is: https://sfu-primo.hosted.exlibrisgroup.com/primo-explore/search?vid=SFUL&sortby=rank.

2.3 From Community Sustainability to Community Productivity

Sustainable development (SD) emerged after the 1987 Brundtland Commission report showed the interconnectedness between human activities and increasing environmental degradation: 26% of the world's population living in developed countries consumed 80–86% of non-renewable resources and 34–53% of food products (WCED, 1987). SD has since represented a new way of thinking about economic development: *doing development differently* (Roseland, 2012). This view is reflected in important international agreements, such as the UN global development agenda for 2030, the 2015 UN (Paris) Climate Change agreement, and the UN New Urban Agenda (United Nations, 2015a, 2015b, 2017).

Our research focus is sustainability at the community level, particularly the complex, adaptive, and interconnected urban systems that require interdisciplinary study (Uphoff, 2014). An urban area is "a human settlement characterized—ecologically, economically, politically and culturally—by a significant infrastructural base; a high density of population, whether it be as denizens, working people, or transitory visitors; and what is perceived to be a large proportion of constructed surface area relative to the rest of the region"(James, 2015).

The UN Global Agenda for 2030 (Sustainable Development Goals) includes a goal for *inclusive, safe, resilient, and sustainable* cities, since cities occupy 3–4% of the world's land surface, use 80% of resources, discharge most global waste, are increasingly vulnerable to climate change impact and health challenges such as depression, and will be host to two thirds of the world's population by 2050 (Girardet, 2015; UN DESA, 2018). The growing awareness that achieving sustainability requires societal change through collaborative decision-making and community engagement has brought sustainable community development (SCD) to the foreground (Clarke, 2012; Hermans et al., 2011).

SCD is a holistic approach that integrates social, environmental, and economic considerations into the dynamic processes and actions of communities on their path toward sustainability, while providing for current and future generations (Berke & Conroy, 2000; Roseland, 2012). In its early steps, SCD was mostly influenced by weak sustainability approaches that champion human domination over nature and infinite economic growth and advise cities to increase their economic output through technology, innovation, and capital accumulation (Ayres, 2007; Solow, 1993; Williams & Millington, 2004).

However, SCD theory has been gradually moving toward a stronger sustainability model which acknowledges the finite character of natural resources and the need for socio-ecological and economic resilience, thus encouraging cities to reduce their resource extraction and consumption (Meerow et al., 2016; Williams & Millington, 2004). SCD has in recent years expanded its scope to embrace stronger sustainability concepts such as social, green, and circular economy, just and collective action, local resilience, and self-reliance (Agyeman, 2008; Connelly et al., 2013; Folke, 2006; Jackson & Victor, 2011; Robinson & Cole, 2015). Despite the conceptual evolution though, community sustainability policies and initiatives have not always involved a balanced application of environmental, economic, and social principles, as reflected in the variety of local agendas and tools grounded in diverse theoretical backgrounds and frequently reflecting specific stakeholders' interests (Joss et al., 2015; Roseland & Spiliotopoulou, 2017).

In pursuit of their well-being, communities are challenged by the difficulties of addressing multiple objectives, thinking strategically and holistically about high-level goals, meaningfully engaging their citizens, and assessing progress consistently (Caprotti et al., 2017; Connelly et al., 2013). SCD requires fundamental changes to the status quo to stop *sustaining* an ill-functioning and unsustainable system with business-as-usual operations driven by growth, in favour of achieving meaningful improvements to community health, happiness, and well-being (Roseland & Spiliotopoulou, 2017).

We suggest that a way to address a wide range of current community sustainability issues is through a transition from the current demand-driven, resource-extracting model to a systemic, resource-productive model of a sustainable city. This transition involves shifting community development from a negative individualistic logic (reducing impact) to a positive systemic one (regeneration of all types of resources within a network of systems) so that the system we *sustain* thereafter is a well-functioning one (Brugmann, 2015; Girardet, 2015).

During this shift to systemic development, community, people, and environment would be involved in a co-evolutionary process, engaging all related systems, subsystems, and stakeholders (Neuman, 2005). Communities that are planned using strong sustainability principles can lead to increases in human, social, resource, and process productivity; improved urban assets performance and systemic interactions; ecological function regeneration; efficient use of resources; and ultimately happiness and well-being (Brugmann, 2015; Girardet, 2015). In the next section we will explain how looking at community development through the conceptual lens of productivity can provide both a bridge linking personal and collective development, and tools to benefit planning for and achieving community happiness and well-being.

2.4 Happiness and Well-Being in a Productive Community

2.4.1 The Interdisciplinary Concept of Community Productivity

Community development—particularly in urban settings—has for a long time focused on reducing impact and risk; for instance, the origins of urban sprawl date to when people needed to address serious health and safety risks and polluted city centres (Neuman, 2005). Today, this focus should be shifting to the development of sustainable and resilient communities which still need to address problems such as growing rates of urbanisation, environmental degradation, extensive resource use, and decreasing subjective well-being. Cultivating, attaining, and sustaining well-being is a process with many parallels to SCD and community productivity, an emerging SCD concept that is forward-looking, acknowledges the systemic nature of the community as a living system, and, in addition to solving problems, focuses on achieving positive outcomes for individual and community well-being.

Conceptually, holistic community productivity is informed by numerous theories and approaches and is multi-dimensional and interdisciplinary, as we will present in this section while linking the approaches and dimensions with community happiness and well-being. The community productivity approach is grounded in strong sustainability principles and seeks maximisation and regeneration of the various forms of community capital. These forms represent the tangible and intangible assets and aspects of a community beyond the traditional triple-bottom line of SD. SCD incorporates natural, physical, economic, human, social, and cultural dimensions of community development (Roseland, 2012) and happiness is part of what the World Bank calls *intangible capital residual* or *intangible assets* (Hamilton & Liu, 2013).

The concept of productivity is usually associated with economic activities and there is extensive research on ***economic and labour productivity***. For instance, this type of productivity is higher in cities that attract agglomeration economies and high-skilled employees—in developed and developing countries alike (Abel et al., 2012; Behrens et al., 2015; Glaeser & Xiong, 2017). Higher labour productivity need not mean increased working hours, though, as it is well established that longer work hours are associated with lower level of happiness and life satisfaction. On the contrary, life-work balance that embraces *time affluence* (free, quality time for oneself and family) contributes to long-lasting positive experiences including happiness (Knight et al., 2013).

Resource productivity and its circularity aspect can contribute to higher levels of happiness and well-being by developing communities that are as self-reliant, locally connected, and resilient as possible. Community productivity also adopts the concept of urban metabolism to analyse the urban ecosystem by studying resource flows and promoting effective policies for a cradle-to-cradle approach (Beloin-Saint-Pierre et al., 2017). In a productive community, circular economy initiatives would not only urge resource regeneration and closing technical cycles in production and consumption, but they would also contribute to the recovery and restoration of the natural environment and improve social and human well-being (Ellen MacArthur Foundation, 2017; Geissdoerfer et al., 2017; World Economic Forum, 2018).

Ecological productivity, or the restoration, enhancement, and regeneration of ecological diversity in an urban area and in its hinterland, can add significant value to the entire local living system, including its human residents. One way of increasing ecological productivity is through regenerative or net-positive design, a promising practice rooted in ecology and living systems theory (Robinson & Cole, 2015) that has been applied to agricultural (Rodale Institute, 2014) and architectural practices (Thomson & Newman, 2018). Generally, the regenerative or productive sustainability principles provide the urban environment designers with a holistic socio-ecological systems worldview that includes collaborative planning and participatory backcasting and is aligned and synergistic with the natural environment (Mang et al., 2016; Robinson & Cole, 2015).

In terms of ***social productivity***, if coupled with a diverse and innovative economy, fair and productive labour, and ecologically enhanced and diverse urban environment, the cultivation of human and social capital is a necessary condition to achieve community happiness and well-being. Many cities today tend to resemble *social deserts*, despite the potential existence of parks and recreation space, or a patchwork of *food deserts*, since people usually need to drive to the nearest grocery store. While it is crucial to find remedies for these *deserts*, social productivity also advocates for the right to retreat as explained for instance by Montgomery (2013): social geometry refers to the optimum setback or yard depth is 10.6 ft or 3.23 m because it allows for conversation and interaction with neighbours or passersby while protecting individual and family privacy.

The definition of ***happiness*** has evolved along with the evolution and dictates of consumerism in the twentieth century, but studies show that, after a certain income level, the increases in economic well-being may have little positive impact on overall well-being and people's subjective perception of happiness (Afsa et al., 2009; Sharpe, 2002). The findings of a large-scale study (sample of 1.7 million individuals worldwide) on correlations between income and life satisfaction (among other factors) agree with previous studies: beyond a specific threshold of income (which varies between different regions of the world), there does not seem to be any strong correlation between income and happiness (Jebb et al., 2018).

Happiness therefore is a function of several factors, aside from the strong determinant of genetic predisposition for happiness or unhappiness and beyond that level of income that satisfies people's basic needs (Frey, 2018). Some factors depend on individual actions and choices (e.g. education, health, work contentment, time affluence, work-life balance, spirituality, etc.) whereas others relate to community elements (e.g. tightness of social connections, community belonging, walkability, access to food, access to recreation and other public amenities, safety, democratic processes, institutional trust, volunteering, civic empowerment, etc.) (Agyeman, 2013; Anielski, 2007; Gilding, 2012).

By enhancing social capital, the productive city becomes an enabler of happiness, reversing current trends of social isolation, loneliness and fear, and providing opportunities for human connections; because social capital can deplete only if it is not used or *produced* (Roseland, 2012). In a socially productive community, the convergence of happiness, well-being, and sustainability is possible through a focus on non-material wealth—or intangible forms of individual and community assets (O'Brien, 2013). Recent research on the urban commons highlights the importance of regenerating and co-producing community capital in an equitable, accessible, inclusive, and creative way; in a way that invites community members not only to share space but also to *co-produce* their well-being through ecologically responsible paths (Elmqvist et al., 2019; Mclaren & Agyeman, 2017; Wahl, 2016).

2.4.2 *Evaluating Well-Being in a Productive Community*

Communities develop and implement sustainability plans often with the aid of a framework or tool; there are many such tools available today and most are designed to support decision-making at all stages, from planning and implementation to monitoring and evaluation (Jong et al., 2015). The community productivity approach can contribute to measuring and evaluating a community's progress towards achieving sustainability goals, by using objective and subjective information guided by a holistic perspective. A community indicators dashboard is often a collection of indicators and does not necessarily provide a coherent or comprehensive picture of the community's health. We should recognise though that it is a snapshot in time, of a dynamic system, and that context, place, and the interconnections of the system's components are significant but intangible parameters that cannot always be captured by static metrics.

Although we usually measure problems such as depression, crime, inaccessibility and social exclusion more than we measure goals such as happiness and well-being, there are initiatives that attempt to evaluate these more intangible qualities globally; for instance, social trust is now included in the World Bank's calculations of the total wealth of nations. Costanza (2014) lists 14 alternatives to GDP to measure well-being, such as the Canadian Index of Well-being, the World Values Survey, Gross National Happiness, Human Development Index, Happy Planet Index, etc. and there are many more nationally or internationally, for instance the European Social Survey, the Canadian General Social Survey, and Vital Signs from Community Foundations of Canada. Most have been developed for use at the national level but can be adapted to the local level since they use either surveys for subjective well-being or proxy metrics. The Seattle Area Happiness Initiative and Sydney's Community Wellbeing Indicators project provide two examples of locally-relevant and meaningful indicators systems (City of Sydney, n.d.; Musikanski et al., 2017).

A productive community can be measured using a combination of effective and widely-accepted indicators of sustainability and several new indicators specifically geared toward the productive and regenerative aspects of the community. Some of these indicators would be considered as proxies for qualitative information around happiness, trust, human connection, governance, and other components of human and social capital. These are some examples of community productivity indicators that are directly or indirectly relevant to happiness and well-being (Spiliotopoulou, 2021):

- Growing space per dwelling unit (proxy for community gardens)
- Mix of land uses and compact development that minimizes commuting times
- Regeneratively designed buildings (net-positive, energy label homes)
- Local innovation (patents)
- Creative industry jobs
- Work opportunities for people with developmental disabilities
- Lifelong learning (vocational or other adult education opportunities)
- Positive individual health practices

- Life satisfaction and/or happiness perception
- Number of neighbours one can rely on in an emergency
- Perceptions of optimism and mental health
- Social service volunteering
- Confidence and trust in local government
- Participation in local/neighbourhood events
- Healthy and safe neighbourhood development initiatives
- Cultural access and participation
- Investment in public art and public art awareness.

During our research, we identified various initiatives of community productivity around the world that promote community happiness and well-being, directly or indirectly. These are some examples of productive or regenerative community practices (8 80 Cities, n.d.; Girardet, 2015; Mang et al., 2016; Razavi, 2017, 2018; Roseland, 2012; Scharmer, 2018; Urban Innovation Community, 2015; Wahl, 2016):

- Adelaide, Australia (dynamic public consultations, major organic waste composting schemes, impressive renewable energy development),
- Copenhagen, Denmark (successful co-housing and community-building projects, energy efficiency initiatives, public transit and cycling uptake, extensive information campaigns and debates, exemplary waste management),
- Amsterdam, The Netherlands (successful sharing and collaborative economy ventures),
- Bristol, U.K. (renewable energy initiatives, successful civil society partnerships and climate resilience actions),
- Medellin, Colombia (inclusive social practices, long-term participatory planning, efficient transportation system),
- Kigali, Rwanda (Africa's leader in knowledge-based sharing economy),
- San Francisco, USA (Quesada Gardens—a regenerative neighbourhood and social capital-building initiative),
- Portland, USA (Eco-District initiative),
- Theory U Lab at MIT (personal and skills development research and application),
- *8 80 Cities* organisation, Toronto, Canada (safety and well-being initiatives)
- Havana, Cuba, New York City, Shanghai, and other places (urban farming and regenerative agriculture programmes).

2.5 Discussion

A happy city is a social and trusting city, one whose social, human, and cultural dimensions of capital are co-produced by all its components. Cultivating a sense of place is a fundamental component of healthy and happy communities. As Orr (2013) explains, looking at the world through the lens of place promotes a sense of responsibility for and a sense of unity with the natural environment. Reclaiming the urban commons and developing a sense of community can help build social capital through place-making and co-production of public space, sharing common places, and fulfilling the right to the city for every person and for nature and biodiversity; thus creating a sense of sustainable happiness as well (Mclaren & Agyeman, 2017; Montgomery, 2013; O'Brien, 2013).

By seeing the community as a dynamic organism and by making the commons (built or natural) more open, creative, and attractive, these spaces become accessible, inclusive, and inviting to produce social connections and sense of belonging for multiple groups (Helliwell et al., 2014; Landry, 2008; Smithsimon, 2008; Wahl, 2016). This approach can also be helpful in times when the community's resilience is tested: increased social resilience, which involves high levels of social trust, institutional trust, neighbourhood connections, happy and resilient individuals, etc., has contributed to tight support networks and better recovery management after a disaster or a socio-economic crisis (Helliwell et al., 2014; Yamamura et al., 2015).

The principles and components of community productivity as outlined in this essay can contribute to effectively address community sustainability planning and implementation through systemic and long-term thinking, equity and justice, urban co-production and governance, and regeneration of urban resources (Condon, 2019; Elmqvist et al., 2019). The concept has the potential to help decouple happiness and well-being from growth and income while promoting sustainable and equitable production and consumption of tangible and intangible forms of community assets (Girardet, 2015; O'Brien, 2013; Roseland & Spiliotopoulou, 2017).

2.6 Conclusion

The significance of producing and regenerating all forms of community capital is increasingly being recognized by SCD researchers and practitioners. Community development should be guided by approaches that promote strong sustainability values and whole-systems thinking that help prevent people from *bowling alone* (Putnam, 1995). We must address this century's realities and planetary constraints with a multi-level systems approach that promotes co-creation of social capital, communities developed with nature-based and regenerative design, and decreased consumption of non-renewable resources—all without compromising collective and individual health and well-being.

Although productivity is historically associated with economic and labor resources, holistic community productivity is interdisciplinary, multi-dimensional, and grounded in strong sustainability. It has been informed by theories and approaches that promote resource circularity and regeneration; restoration of the health of socio-ecological systems; socio-cultural equity, inclusion, happiness, and well-being; and collaborative and inclusive partnerships for community co-creation. The application of community productivity principles, under the umbrella of systems thinking, can help increase quality of life despite and perhaps in part because of lower levels of material expectations (*standard of living*) by increasing well-being and happiness.

Productivity acknowledges the interdependence of all enablers of happiness, individual and community alike, and aspires to guide all actors, designers, and co-producers of happy communities. The community productivity concept has real potential to allow community elements to converge and move forward in a balanced and synergistic way, and with a common goal in mind: to achieve a happy, resilient, healthy, smart, creative, livable, cooperative, just, inclusive, and sustainable community.

References

80 Cities. (n.d.). *8 80 Cities*. https://www.880cities.org/

Abel, J. R., Dey, I., & Gabe, T. M. (2012). Productivity and the density of human capital. *Journal of Regional Science, 52*(4), 562–586. https://doi.org/10.1111/j.1467-9787.2011.00742.x

Afsa, C., Blanchet, D., Marcus, V., Pionnier, P. A., Rioux, L., D'Ercole, M. M., Ranuzzi, G., & Schreyer, P. (2009). *Survey of existing approaches to measuring economic performance and social progress*. INSEE.

Agyeman, J. (2008). Toward a "just" sustainability? *Continuum, 22*(6), 751–756. https://doi.org/10.1080/10304310802452487

Agyeman, J. (2013). *Introducing just sustainabilities: Policy, planning, and practice*. Zed Books. https://www.zedbooks.net/shop/book/introducing-just-sustainabilities/

Anielski, M. (2007). *The economics of happiness: Building genuine wealth*. New Society Publishers.

Ayres, R. U. (2007). On the practical limits to substitution. *Ecological Economics, 61*(1), 115–128. https://doi.org/10.1016/j.ecolecon.2006.02.011

Behrens, K., Duranton, G., & Robert-Nicoud, F. (2015). Productive cities: Sorting, selection, and agglomeration. *Journal of Political Economy, 122*(3), 507–553.

Beloin-Saint-Pierre, D., Rugani, B., Lasvaux, S., Mailhac, A., Popovici, E., Sibiude, G., Benetto, E., & Schiopu, N. (2017). A review of urban metabolism studies to identify key methodological choices for future harmonization and implementation. *Journal of Cleaner Production, 163*, S223–S240. https://doi.org/10.1016/j.jclepro.2016.09.014

Berke, P. R., & Conroy, M. M. (2000). Are we planning for sustainable development? *Journal of the American Planning Association, 66*(December), 21–33. https://doi.org/10.1080/01944360008976081

Brugmann, J. (2015). The urban productivity imperative. *The productive city: Growth in a no-growth world* (pre-publication draft). The Next Practice Ltd.

Caprotti, F., Cowley, R., Datta, A., Broto, V. C., Gao, E., Georgeson, L., Herrick, C., Odendaal, N., & Joss, S. (2017). The new urban agenda: Key opportunities and challenges for policy and practice. *Urban Research & Practice, 10*(3), 367–378. https://doi.org/10.1080/17535069.2016.1275618

City of Sydney. (n.d.). *City of Sydney's community wellbeing indicators*. https://www.cityofsydney.nsw.gov.au/learn/research-and-statistics/community-indicators

Clarke, A. (2012). *Green municipal fund. Passing go: Moving beyond the plan*. Federation of Canadian Municipalities.

Condon, P. M. (2019). *Five rules for tomorrow's cities: Design in an age of urban migration, demographic change, and a disappearing middle class*. Island Press.

Connelly, S., Markey, S., & Roseland, M. (2013). We know enough: Achieving action through the convergence of sustainable community development and the social economy. In R. Simpson, & M. Zimmerman (Eds.), *The economy of green cities* (pp. 191–203). Springer. https://doi.org/10.1007/978-94-007-1969-9

Costanza, R. (2014). Time to leave GDP behind. *Nature, 505*, 283–285. https://doi.org/10.1038/505283a

De Jong, M., Joss, S., Schraven, D., Zhan, C., & Weijnen, M. (2015). Sustainable-smart-resilient-low carbon-eco-knowledge cities; making sense of a multitude of concepts promoting sustainable urbanization. *Journal of Cleaner Production, 109*, 25–38. https://doi.org/10.1016/j.jclepro.2015.02.004

Ellen MacArthur Foundation. (2017). *Cities in the circular economy: An initial exploration.*

Elmqvist, T., Andersson, E., Frantzeskaki, N., McPhearson, T., Olsson, P., Gaffney, O., Takeuchi, K., & Folke, C. (2019). Sustainability and resilience for transformation in the urban century. *Nature Sustainability, 2*(4), 267–273. https://doi.org/10.1038/s41893-019-0250-1

Folke, C. (2006). Resilience: The emergence of a perspective for social-ecological systems analyses. *Global Environmental Change, 16*(3), 253–267. https://doi.org/10.1016/j.gloenvcha.2006.04.002

Frey, B. S. (2018). *Economics of happiness.* Springer International Publishing AG.

Geissdoerfer, M., Savaget, P., Bocken, N. M. P., & Hultink, E. J. (2017). The circular economy—A new sustainability paradigm? *Journal of Cleaner Production, 143*, 757–768. https://doi.org/10.1016/j.jclepro.2016.12.048

Gilding, P. (2012). The great disruption: Why the climate crisis will bring on the end of shopping and the birth of a new world. *Bloomsbury Press.* https://doi.org/10.1177/0973408213475379

Girardet, H. (2015). *Creating regenerative cities.* Routledge.

Glaeser, E. L., & Xiong, W. (2017). Urban productivity in the developing world. *Oxford Review of Economic Policy, 33*(3), 373–404. https://doi.org/10.1093/oxrep/grx034

Hamilton, K., & Liu, G. (2013, March). *Human capital, tangible wealth, and the intangible capital residual.* Policy research working paper WPS 6391.

Helliwell, J. F., Huang, H., & Wang, S. (2014). Social capital and well-being in times of crisis. *Journal of Happiness Studies, 15*(1), 145–162. https://doi.org/10.1007/s10902-013-9441-z

Hermans, F. L. P., Haarmann, W. M. F., & Dagevos, J. (2011). Evaluation of stakeholder participation in monitoring regional sustainable development. *Regional Environmental Change, 11*(4), 805–815. https://doi.org/10.1007/s10113-011-0216-y

Jackson, T., & Victor, P. (2011). Productivity and work in the "green economy." *Environmental Innovation and Societal Transitions, 1*(1), 101–108. https://doi.org/10.1016/j.eist.2011.04.005

James, P. (2015). *Urban sustainability in theory and practice: Circles of sustainability.* Earthscan Publications Ltd.

Jebb, A. T., Tay, L., Diener, E., & Oishi, S. (2018). Happiness, income satiation and turning points around the world. *Nature Human Behaviour, 2*, 33–38.

Joss, S., Cowley, R., De Jong, M., Müller, B., Park, B. S., Rees, W. E., Roseland, M., & Rydin, Y. (2015). *Tomorrow's city today: Prospects for standardising sustainable urban development.* University of Westminster.

Knight, K. W., Rosa, E. A., & Schor, J. B. (2013). Could working less reduce pressures on the environment? A cross-national panel analysis of OECD countries, 1970–2007. *Global Environmental Change, 23*(4), 691–700. https://doi.org/10.1016/j.gloenvcha.2013.02.017

Landry, C. (2008). *The creative city: A toolkit for urban innovators.* Routledge.

Mang, P., Haggard, B., & Regenesis. (2016). *Regenerative development and design: A framework for evolving sustainability.* Wiley.

Mclaren, D., & Agyeman, J. (2017). *Sharing cities. Environment: science and policy for sustainable development* (Vol .59, Issue 3, pp. 22–27). https://doi.org/10.1080/00139157.2017.1301168

Meerow, S., Newell, J. P., & Stults, M. (2016). Defining urban resilience: A review. *Landscape and Urban Planning, 147*, 38–49. https://doi.org/10.1016/j.landurbplan.2015.11.011

Montgomery, C. (2013). *Happy city.* Random House Canada.

Musikanski, L., Polley, C., Cloutier, S., Berejnoi, E., & Colbert, J. (2017). Happiness in communities: How neighborhoods, cities, and states use subjective well-being metrics. *Journal of Social Change, 9*(1), 32–54. https://doi.org/10.5590/JOSC.2017.09.1.03

Neuman, M. (2005). The compact city fallacy. *Journal of Planning Education and Research, 25*(1), 11–26. https://doi.org/10.1177/0739456X04270466

O'Brien, C. (2013). Happiness and sustainability together at last! Sustainable Happiness. *Canadian Journal of Education, 36*(4), 228–256.
Orr, D. W. (2013). Place and pedagogy. *The NAMTA Journal, 38*(1), 183–188. https://files.eric.ed.gov/fulltext/EJ1078034.pdf
Putnam, R. D. (1995). Bowling alone: America's declining social capital. *Journal of Democracy, 6*(1), 65–78. https://doi.org/10.1353/jod.1995.0002
Razavi, L. (2017). *Building a city: Regulating the sharing economy in Amsterdam*. Medium. https://medium.com/@LaurenRazavi/building-a-city-regulating-the-sharing-economy-in-amsterdam-faecee8dfb0
Razavi, L. (2018). *How Rwanda's capital became an African tech leader*. Medium. https://medium.com/s/story/how-rwandas-capital-became-an-african-tech-leader-66e00edc74d
Robinson, J., & Cole, R. J. (2015). Theoretical underpinnings of regenerative sustainability. *Building Research and Information, 43*(2), 133–143. https://doi.org/10.1080/09613218.2014.979082
Rodale Institute. (2014). *Regenerative organic agriculture and climate change*. White paper.
Roseland, M. (2012). *Toward sustainable communities: Solutions for citizens and their governments* (4th ed.). New Society Publishers.
Roseland, M., & Spiliotopoulou, M. (2017). Sustainable community planning and development. In M. A. Abraham (Ed.), *Encyclopedia of sustainable technologies* (Vol. 2). Elsevier. https://doi.org/10.1016/B978-0-12-409548-9.10185-X
Scharmer, C. O. (2018). *The essentials of theory U : core principles and applications*. Berrett-Koehler Publishers, Inc. https://www.penguinrandomhouse.com/books/565259/the-essentials-of-theory-u-by-c-otto-scharmer-and-katrin-kaufer/9781523094400/
Sharpe, A. (2002). Productivity concepts and trends. *The review of economic performance and social progress* (Vol. 3, Issue 33, pp. 29–56). http://www.csls.ca/repsp/2/andrewsharpe.pdf
Smithsimon, G. (2008). Dispersing the crowd: Bonus plazas and the creation of public space. *Urban Affairs Review, 43*(3), 325–351. https://doi.org/10.1177/1078087407306325
Spiliotopoulou, M. (2021). *Sustainable community development through the conceptual lens of productivity*. Unpublished doctoral dissertation. Simon Fraser University.
Solow, R. M. (1993). An almost practical step toward sustainability. *Resources Policy, 19*(3), 162–172. https://doi.org/10.1016/0301-4207(93)90001-4
Thomson, G., & Newman, P. (2018). Urban fabrics and urban metabolism—from sustainable to regenerative cities. *Resources, Conservation and Recycling, 132*, 218–229. https://doi.org/10.1016/j.resconrec.2017.01.010
UN DESA. (2018). United Nations. World urbanization prospects: The 2018 revision [Key facts]. *Economic and Social Affairs*. https://population.un.org/wup/Publications/Files/WUP2018-KeyFacts.pdf
United Nations. (2015a). Adoption of the Paris Agreement (FCCC/CP/2015/L.9/Rev.1). http://unfccc.int/resource/docs/2015/cop21/eng/l09r01.pdf
United Nations. (2015b). Resolution 70/1. Transforming our world: the 2030 Agenda for Sustainable Development. https://sustainabledevelopment.un.org/post2015/transformingourworld
United Nations. (2017). *New Urban agenda*. In Conference on housing and sustainable urban development (Habitat III). ISBN: 978-92-1-132757-1.
Uphoff, N. (2014). Systems thinking on intensification and sustainability: Systems boundaries, processes and dimensions. *Current Opinion in Environmental Sustainability, 8*, 89–100. https://doi.org/10.1016/j.cosust.2014.10.010
Urban Innovation Community. (2015). *Telling city success stories—Publication of roundtable on SDGs and urban innovation*.
Wahl, D. C. (2016). *Designing regenerative cultures*. Triarchy Press.
WCED. (1987). *Report of the World Commission on environment and development: Our common future*. The Brundtland Report. https://doi.org/10.1080/07488008808408783
Williams, C. C., & Millington, A. C. (2004). The diverse and contested meanings of sustainable development. *The Geographical Journal, 170*(2), 99–104. https://doi.org/10.2307/3451586

World Economic Forum. (2018). *Circular economy in cities: Evolving the model for a sustainable urban future*. White paper. https://www.weforum.org/whitepapers/circular-economy-in-cities-evolving-the-model-for-a-sustainable-urban-future

Yamamura, E., Tsutsui, Y., Yamane, C., Yamane, S., & Powdthavee, N. (2015). Trust and happiness: Comparative study before and after the great East Japan earthquake. *Social Indicators Research, 123*(3), 919–935. https://doi.org/10.1007/s11205-014-0767-7

Chapter 3
Sustainability and Happiness as "Promoting Life"

Jacob Bethem

Abstract While various authors have tried to merge happiness and sustainability together in such concepts as *sustainable happiness* and *sustainable well-being*, among others, the link is tenuous as the theories built around these concepts are incomplete. In this work, I strive to fill this gap by integrating further conceptualizations of the *value of life*. While happiness and sustainability are not one in the same, since someone can be happy in unsustainable ways and since something can be performed sustainably without causing happiness, I reflect on the overlap where happiness and sustainability come together. This overlap seems to be the most meaningful and purposeful notions of the terms and the sense in which they have their most normative force. I will attempt to identify the overlap to define both happiness and sustainability as *the promotion of life*. I will evaluate whether and how this stipulative, working definition for these terms can be useful for practical guidance in real projects in an attempt to bridge theory and application. I review the place of happiness and positive psychology in well-being philosophy and recent advances incorporating sustainability into well-being theory. I also discuss various conceptions of the value of life. I then incorporate these preceding philosophies together to form a conceptual map of the variety of assessments of life that exist. From these concepts, I develop preliminary principles that might serve as a guide for practical living, for happier, more sustainable, and more ethical lives.

Keywords Happiness · Sustainability · Good life · Vitalism · Well-being · Love

3.1 Introduction

I began this project wondering whether the term *the good life*—the philosophical notion that could broadly describe a life well-lived or more narrowly refer to an ideal or the ultimately best life, could be better understood by dissecting its component

---

J. Bethem (✉)
Department of Business, Economics and Enterprise Sciences, California University of Pennsylvania, California, Pennsylvania, USA
e-mail: bethem@calu.edu

S. Cloutier et al. (eds.), *Linking Sustainability and Happiness*,
Community Quality-of-Life and Well-Being,
https://doi.org/10.1007/978-3-030-89559-4_3

terms. That is, the good life probably depends on *life*, but how and in what sense was not obvious. Asking what makes a life good, I wondered whether that was the same as asking what a life lived well is. I was immediately reminded of the philosophy about well-being, which could give further clues. According to ethicists, well-being is doing well and being ethical, but I wondered if doing well is necessary to the concept of a good life, or if being ethical is all that matters. Ethicists are often criticized for including ethical requirements within well-being, but I wondered if not only are they right that ethics should be included, but there was a possibility that perhaps the other components were extraneous. Upon further reflection of what it means to be ethical, I was reminded of Schweitzer (1936), a renowned ethicist who contends that ethics is ultimately about affirming life. So, at this point, I conjectured that the good life could be related to living well, or well-being, which could be related to ethics, which might be (or mean) affirming life. I had a few interrelated terms that fundamentally described living well as affirming life. So, the suggestion I reached was that possibly the good life is about affirming life. If *to affirm* something is to find it good, then to live the good life is to find life good. At this point, I had to wonder whether this reasoning reflected a circular argument or was something more profound.

Because of my educational background, at the same time I was wondering how this inquiry relates to sustainability. Sustainability concerns conservatively using resources, the triple bottom line, and sustaining life. In the same way that I questioned whether well-being needed the other components besides ethics, I wondered whether the other aspects of sustainability are extraneous or perhaps already captured in the concern for sustaining life. That is, is sustainability fundamentally about just sustaining life? If so, sustainability could perhaps be said to be life-affirming, just like ethics and the good life. Then, sustainability might also be related to the good life, or sustainability might *be* the good life. Now, I once again had to question whether I made any progress in understanding these terms or if I was just conflating them.

In this chapter, I will review terms associated with the good life in order to try to better understand it. I begin by reviewing *well-being* and *happiness* because these terms are probably the most used in the philosophy of the good life. Next, I address sustainability because there is a growing contention that this concept embodies the same notion. To keep the conversation focused, I will attempt to continuously return the discussion back to *life* as a central focus. To enhance that focus, I specifically investigate the definition of *life* and its value in extended sections of the chapter in a deeper dive. The central question of this chapter is: "What can sustainability and happiness reveal about the good life?" My thesis is that investigating concepts such as sustainability and happiness in terms of what they reveal about life, its goodness, and its value provides meaningful guidance for how to live the good life.

3.2 Methodology

To investigate the central research question, I will use methods common to philosophical inquiry such as leveraging healthy skepticism or casting doubt, critiquing ideas and questioning beliefs, and attempting logical argumentation with counterarguments. In other words, I will play the role of devil's advocate at times rather than merely following my pre-established beliefs or arguing a predestined conclusion. Some portions will reflect the three philosophical approaches described by Kienstra et al. (2014)—connective truth finding, truth testing, and judgments. It might be better to call the proceeding sections *language games* after Wittgenstein's (1953) reflections that suggest that much philosophical inquiry is about finding the right words. I will be interrelating many concepts. I will establish a foundation with conventional theories from the literature and interrelate them with more recent research. However, I also hope to build on them using further inquiries in new directions to provide novel insights. I will not have the space to challenge every thought presented but, hopefully, I will give objectively fair scrutiny. Admittedly, the path of inquiry is guided to some extent by my curiosity, and some of my personal reactions may unintentionally slip into the discussion. Still, the investigation is self-serving, but I publish it here in case it can be helpful to others. I have attempted to remove less insightful tangents. This work is more speculative than definitive, and an open debate is invited.

3.3 Conceptual Review

Philosophers, particularly ethicists, are concerned with finding *the good life.*[1] At least two (potentially synonymous[2]) concepts are associated with the good life—*well-being* and *happiness*. In recent decades, growing concern has arisen about how lifestyle choices and the societal institutions that enable them or are enabled by them have created an existential threat related to climate change. So, this high-priority issue of sustainability studies (i.e., how to live in such a way as to avoid dangers related to climate change) is becoming more important than ever in investigations of the good life. I will review some of the overlaps among these concepts to see what more it can tell us about the good life.

3.3.1 Well-Being

Well-being generally refers to "how (a) life is going."[3] In a technical review of the concept, Roger Crisp (2017) utilizes the same taxonomy created by Derek Parfit

---

[1] For purposes here, I will alternate between articles *a* and *the* in front this term. I do not mean to definitively declare that there is one or are multiple lives that could be *good life* in the *big question* sense that seems to suggest that that life is the best (possible) life.

[2] They may be potentially synonymous with each other or with the good life.

[3] I intentionally put the article in parentheses to indicate that the life being evaluated might be singular (an individual), plural (a group), or a hybrid (some mix or relationship between individuals and groups.

(1984), which describes three conceptualizations of well-being.[4] While the proponents of these arguments often debate which of these theories is the correct one, because of their weaknesses, for the purposes of this chapter, I will treat well-being as a multi-dimensional term and each of these theories about it as describing a component of well-being, rather than arguing between them. That is, a life would be going better if it had each of these aspects than if it had less of or was missing them. I will treat well-being as something that comes in degrees, such that a life can be going more or less well than at other times or in comparison to other lives, rather than as a binary concept that would describe life as either well or not.

The three perspectives, which I will treat as components of well-being, are hedonism, desire satisfaction, and objective lists. Hedonistic theories of well-being describe life as going well if it is pleasurable and avoiding pain. However, endless pursuit of pleasure is questioned through examples like Nozick's (1974) *experience machine*, an imaginary device that could feed endless pleasurable stimulations to your brain as if you were actually experiencing them. The undesirability of such a device suggests that there may be more to a good life than just the experience of pleasure, such as actual pleasure, but pleasure remains a salient indicator of a good life.

Desire-based theories of well-being describe a life as going well if its preferences are being satisfied. These preferences may have to be rational, present, informed, and global (maybe restricted by further conditions still), but meeting desires remains a salient indicator of a good life, while unmet needs (and possibly desires) might indicate a life that is not going as well.

So-called *objective list* theories of well-being describe a life as going well if specified goods are present in that life, such as knowledge, friendship, and happiness. These lists can be criticized for tending to suspiciously reflect the preferences or values of their authors, making it questionable whether they are created objectively and worthy of the name, even if meant to be evaluated from an impersonal stance. While no list has been finalized and might not be universal (perhaps not everyone needs all goods or all of them to the same degree), having such aspects in one's life remains a salient indicator of a good life.

3.3.2 Happiness

Happiness is described via a variety of conceptualizations. Happiness is when something good happens to you or a positive feeling, such as one you might get from good experiences or events happening to you. Haybron's (2020) review of the literature describes happiness in three ways—as an affective state, as life satisfaction, or as synonymous with well-being (in any of the three forms previously discussed by Crisp).

[4] There may be other conceptualizations besides the foundational types listed here. Some of these perspectives have been augmented by modern thinkers, and still others have been created.

The affective state can take two forms—as an emotional state or a psychological experience (i.e., pleasure, the same as hedonism in well-being philosophy[5]). Happiness as an emotion might include long-term contentment or short-term thrill, but sometimes leaves critics wanting more (often, something that makes it conceptually more like well-being).

Happiness conceptualized as life satisfaction is exemplified in the field of positive psychology. Various surveys are conducted to assess whether participants are happy with their lives. Example of these tests include Likert scales (strongly agree to strongly disagree) or the *Satisfaction with Life Scale* (Diener et al., 1985; Lyubomirsky et al., 2005). They generally ask if the participant's life is fulfilling and avoiding negative states or is regrettable. These tests can resemble desire-satisfaction types of well-being or could include questions about aspects within the objective lists for well-being. Yet, life satisfaction may require more than fulfilling desires; so, they are not the same concept. Life satisfaction tests can be manipulative, done incorrectly, or assessing the wrong stuff (i.e., it is a fallible tool, which draws questions of whether it is the test or the test-taker that are the problem).

Conceptualizing happiness as well-being feels too rich for some critics (who often seek something closer to the emotional explanation). Like the hybrid compromise made with well-being in this chapter, it might be best to say for now that each of these conceptualizations has strengths and weaknesses and might each remain a salient indicator of happiness and of a good life.[6] Because disagreement exists whether well-being is a type of happiness or whether happiness is a component of well-being, two conceptual models are shown in Fig. 3.1, depicting the relationship among their components. For what it is worth, I favor thinking of happiness as a component of well-being.

3.3.3 Integration of Ethics and Others

There is still further distinction that may need to be made with the terms *eudaimonia* and *flourishing*, often associated with Aristotle's or other Greek philosophies, but such a task is more than I can accomplish here (as examples, see Nagel, 1972; Ackrill, 1974; Athanassoulis, n.d.). I mention these additional terms to recognize that

[5] The difference between happiness as pleasure and hedonistic sense of pleasure used in well-being might be whether discussing an experience in a moment or the balance of pleasure over pain over a period of time.

[6] Haybron (2020) also identifies additional factors that might be associated with happiness, well-being, or a good life, such as: a sense of meaning or purpose, meaningful or worthwhile activities, engagement in interesting and challenging activities, achieving goals, personal growth, relatedness, supportive relationships, autonomy or control, competence, material and physical security, a positive outlook, religion, good governance, trust, helping others, values, not being unemployed, connection with the natural environment, household income, possession of luxury conveniences, satisfaction with standard of living, and *psychosocial prosperity*, which includes "being treated with respect in the last day, having family and friends to count on, learning something new, doing what they do best, and choosing how their time was spent."

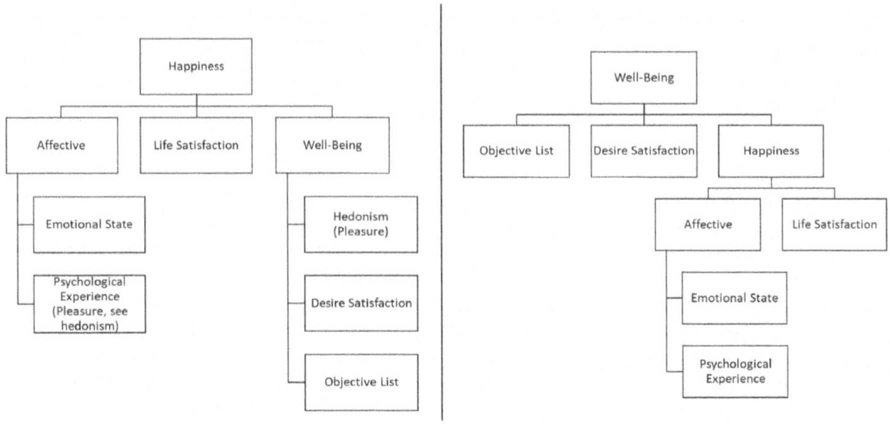

Fig. 3.1 The three components of well-being from Crisp and the three aspects of happiness from Haybron are illustrated. Two ways to portray the conceptual map of the philosophy of well-being and happiness exist because, while often informally treated as synonyms, the technical literature is not yet decided definitively. The first (left) shows well-being as a type of or contributor to happiness. The second (right) shows happiness as a component or determinant of well-being

Aristotle's contribution to the philosophy of happiness, well-being, and the good life includes some intriguing ideas. However, it is not my intention to review Aristotle's philosophy verbatim here. While I will draw some ideas from that philosophy, I will likely take some ideas in directions it might not support. For starters, Aristotle's conception of well-being, as confirmed by Crisp's and Haybron's and in similar fashion to other ethicists, includes both a life that is going well and some element of ethical quality to it. That is, ethicists argue that a life is going well if it is good for the being under consideration and if that being is acting ethically. That is, the good life is not only about being well-off (i.e., doing well), but acting appropriately (i.e., doing right or doing good actions). Ethicists are less likely to attribute the good life to an evildoer. That is, the evildoer's deeds disqualify them from being well-off to most ethicists. A critic might adopt a similar but weaker stance that grants an evildoer some status of well-being but to a lesser degree than someone who shares other characteristics but also acts ethically or might dismiss the ethical criterion as extraneous.

One reason that ethicists might insist on this additional component is because in searching for the good life, they are usually going to prescribe that everyone or as many people as possible ought to live this way. If only some people could live in the prescribed way or if it is at the expense of others (as critics would allow), then it would not be compatible with the universal nature of ethics. That is, this life might be (self-) contradictory or hypocritical for ethicists to prescribe or unethical (in action), or unjust (in result). Still, unless the good life is some type of perfect life, which would make it unattainable by anyone because we are imperfect beings, ethicists might also consider some way of atoning for unethical behavior so that some unethical actions might not disqualify someone from being said to live a good life.

Forgiveness might be an important consideration in this regard (see Ang et al., 2016 for more).

Ethics is the study of right and wrong behavior and is usually associated with actions that impact other people, not just the agent performing the action. When well-being includes an ethical requirement, it suggests that the well-being of others might impact the agent's well-being. If so, it may be the case that if others around us are not doing well (particularly if something we did had something to do with their current state), then we might be said to not be doing as well as we otherwise could be said to be doing. That is, our life is not as good as it otherwise could be simply due to how our life is making others' worse. To the extent that we can separate or distinguish our lives from others, well-being can be an individualized concept; however, if we cannot identify ourselves entirely in this way, then well-being might be a relational or collective concept. There are studies in which people describe their lives as going well or poorly based on their relations or in comparison to others. For instance, someone might be satisfied with their income (on its own) but still might find it inadequate in comparison to others.'

If the *others* with which we identify are numerous, we end up with concepts like *intersubjective well-being* and *community well-being* (see Lee & Kim, 2016). It may be the case that I might not consider myself doing well if my neighborhood, city, state, or country is not doing well. Their issues may weigh on me or reflect unwellness for me. To the extent that I identify as a global citizen, planetary health or the global community's status might also be encompassed or reflected to some extent in whether my life is going well. Adding ethical importance or identifying the life being evaluated across multiple individuals (i.e., a group) makes for a more complex conceptualization.

Although it may be possible to associate a greater context or more lives into the target subject of a well-being evaluation, that does not necessarily mean that such context must be considered. It seems that as long as an audience understands whether the evaluator of well-being is speaking inclusively of the context or of others, that should suffice. For instance, if someone were evaluating the play of a sport team, they can decide whether to include their playoff performance or exclude it (only discussing the regular season). As long as the audience understands the parameters, their evaluation should be clear in that regard. Even a more complicated analysis such as the team's play without the captain, even if the captain played extensively in every game, can be approximated, even though some factors will likely be skewed and difficult to adequately measure without the captain's contributions because they are partially dependent on it. The same might be possible with assessments of well-being. Still, the team's performance with the captain and including the playoffs seems to be a more comprehensive assessment. So, if the analogy holds for well-being assessments, and a more comprehensive assessment is desirable, then the context of others ought to be included in those cases.

3.3.4 Sustainability

Sustainability is an interdisciplinary study of ways to improve how we live and how organizations operate. It is studied across cultures and generations, in innumerable dimensions, but commonly regards such matters as energy, water, waste, health, equity, and biodiversity. Because it is so broad, the term can be used in numerous, sometimes conflicting ways. For instance, a sustainability initiative that is intended to make a factory's operations more sustainable by cutting emissions could mean that the workers have to stop (i.e., not sustain) their current process or livelihoods. In these ways, there might be tradeoffs to consider. While the term *sustain* generally refers to keeping something going or alive, research within the sustainability discipline is often critical of the status quo, as sustainability advocates seek ways to improve upon current practices. In this way, sustainability often is treated as an ethical matter. Proponents sometimes describe situations in which intentionally avoiding a more sustainable behavior is unethical, due to its impacts on or disrespect of others.

The most commonly cited definition of sustainability originated in the United Nations' Brundtland Commission Report, *Our Common Future*: "Sustainable development is development that meets the needs of the present without compromising the ability of future generations to meet their own needs" (United Nations World Commission on Environment & Development, 1987). Elkington's (1998) *triple bottom line*, which might be the second most common depiction of the term, portrays sustainability as a multidimensional interaction between financial, social, and environmental considerations.

Environmental ethicist Jamieson (1998) is among the scholars who have asked such questions as, "Sustainability of what? For whom? Why?" Jamieson answers by citing the work of Rajni Kothari, who recommends, "We should treat all life as indispensable" (Jamieson, 1998, p. 191). Speaking of *needs*, *tradeoffs*, and *values* shows that sustainability concerns ethics. This passage implies that sustainability ethics are founded on the value of *life*. Sustainability is ultimately concerned with continuing life on Earth. Some sustainability advocates are concerned with only sustaining some subset of life, such as human life, a business, or some individual person, but nonetheless, *life* is foundational. Hence, it is easy to see why the existential threat associated with dangerous climate change has become a priority in sustainability studies. Sustainable living generally means to act in such a way as to reduce or reverse your eco-footprint, to act mindfully of others (including nonhumans and future generations), and to use financial and other resources responsibly. Sustainable living includes your personal life, professional life, and public life.

3.3.5 Integration of Sustainability

While various authors have tried to merge happiness and sustainability together in such concepts as *sustainable happiness* and *sustainable well-being*, among others, the link is tenuous at best thus far, because the theories built around these concepts still seem to be missing something.

For instance, O'Brien defines *sustainable happiness* not as "feeling endlessly joyful" but as "happiness that contributes to individual, community and/or global well-being without exploiting other people, the environment or future generations" (pp. 232, 245). This definition mimics the Brundtland Commission definition of sustainability and incorporates a sense of identifying with or relating to others in an ethical sense. O'Brien asserts that well-being goes beyond an individual saying, "our well-being is inextricably associated with the well-being of others and the natural environment," giving an example that implies that happiness experienced as a positive emotion derived from drinking coffee is not sufficient to qualify as sustainable happiness if the coffee "came at the expense of other people or the natural environment" (2013, pp. 232–233, 250). O'Brien's conceptualization of sustainable happiness includes ethical values that encourages avoiding consumerism and isolated individualism in favor of resource conservation and interdependence. In opposition to "a common misconception that living sustainably will lower our quality of life, requires self-sacrifice, or is assumed to detract from happiness," O'Brien contends that "the path to happiness may also be consistent with sustainable lifestyles, can potentially enhance our quality of life, and can contribute to well-being" (O'Brien, 2013, p. 233). O'Brien's perspective could be strengthened by providing an explicit and compelling explanation not just of why an ethical requirement should be included in a conception of well-being, but why these particular ethics related to sustainability studies ought to be included—but her work is less meta-theory and more applied (about activities to enact this view in schools and in policy).[7]

Kjell (2011) argues for a more comprehensive notion of well-being that includes sustainability, taking the individual's social and environmental context into consideration, called *sustainable well-being.* However, whether this more comprehensive term is simply more complex or whether well-being conceptualizations without the addition of sustainability metrics are insufficient to capture the notion is debatable (as it is with the addition of ethical living requirements). Kjell seems to acknowledge that individuals can be isolated from some larger context but sees that as a lesser (not necessarily wrong or false) assessment of well-being than what is provided from something more contextualized. The statement that "life is lived in coexistence and cooperation with others" (2011, p. 261) resembles the philosophical adages that humans are *social* or *relational* creatures. Kjell also acknowledges that the motivation to inspire policies at a macro-level is part of the impetus to conceptualize well-being (and sustainability) in this way (Kjell, 2011, 255), but that "current well-being approaches do not effectively illustrate how to bridge individual with group well-being" (Kjell, 2011, 261). At the policy level, individual well-being assessments are not always as useful as a collective assessment or something with greater context could be. There is an inclination to say that a community is not doing well if it is not adopting practices to improve its sustainability.

[7] O'Brien also uses the terms *health, healthy lifestyle,* and *quality of life,* which may need to be mapped in relation to happiness and well-being.

3.3.6 *Proposing a Thought Experiment*

The variety of terms and conceptualizations described here and the unanswered questions and disputes between them demonstrate that this philosophy is an unsettled matter. I started this discussion asking, "What is the good life?" and identified a variety of concepts that have been discussed as potential answers. However, I wonder whether it was the right question to ask. Other questions I might ask include: What is life? What makes a life good? What is good? Is *good* in good life different than *well* in *well-being*? Also, it has come to light that I need to test whether a life can be going well without being sustainable and vice versa. I will try to find these answers next through a series of thought experiments that illustrate an extended train of thought. I will associate these concepts and provide basic thoughts to critique some of the word associations to see if they provided any further insights into the concept of a good life.

3.4 A (Train of) Thought Experiment(s)

Within this section, I share a series of thought experiments intended to determine where life's goodness or value lies. I will recall four definitions for life, scrutinize how each may be considered good, and compare these ideas to the notions of well-being and sustainability already discussed, leading to more questions regarding life's value and goodness but also a richer mental map.

3.4.1 *Searching for the Meaning of Life*

I want to see what the meaning of life can tell us about the good life. To do that, I will address the *meaning* of life in terms of both its definition and its value.[8] There are at least four ways that we can define *life*:

1. Vitality—being alive
2. Lifestyle—the choices we make, the actions we do
3. Well-Being- all well-being components (i.e., emotional and psychological affective states, desire and life satisfaction, and any aspects within an objective list)
4. Biosphere—the community of all organisms (Bethem, 2018, p. 75).

[8] I cannot fit the sense of the *meaning of life*, meaning purpose that one or any of us is here, exists, or is alive, in the space provided here.

3.4.1.1 Life as *Vitality*

Each definition might offer insight into what makes a good life. First, a good life might mean being alive, and it is common to value longer or more robust lives. So, a life might be good in the sense that the organism can utilize all of its biophysical functions or is fully animated, rather than not fully conscience or capable.[9] Being in good health and acting autonomously might be the ways to demonstrate a good life in this regard. However, longevity is not universally valued, as some critics would prefer not to age or to avoid conditions such as being on life-support (seeing it as a life not worth living). Also, as mortal beings, a good life might necessarily have to include death at some point.[10] Another way to consider vitality as part of a good life could be the perpetuation of offspring, which allows someone in a sense to live-on through their progeny. However, more life in this sense (i.e., more lives, more beings) may not always be better, as it sometimes leads to competitive conflict.

3.4.1.2 Life as *Lifestyle*

Second, a good life might mean *the right lifestyle*, full of good choices or actions. This definition seems to relate the most to the ethics, but some philosophers, such as Little (2014), define well-being as *well-doing*, utilizing a conceptualization of the term focused on the personal projects of the subject. There may be more to the concept than simply doing actions that are ethical on an individual basis. For instance, perhaps those actions also have to conform to a certain life plan, which might be considered ethical on a collective basis.

3.4.1.3 Life as *Well-Being*

Third, a good life might mean the same as well-being, as defined earlier or in some other way. Therefore, the good life might refer to some measure of well-being within the subject being assessed, perhaps optimized, maximally, or complete in the sense of manifesting positively in the multiple dimensions previously described.

3.4.1.4 Life as the *Biosphere*

Fourth, a good life might reflect something good about the interspecies community, perhaps that it is diverse, interrelating well, or reflecting some other positivity. For instance, consider Leopold's (1949) principle that "a thing is right if it tends to

[9] Such a view might have offensive implications for people who are not fully able, but that meaning is not endorsed here.

[10] Due to space, I will not pursue a tangent about *good deaths* here, even though they might be a necessary component of a good life.

preserve the integrity, stability, and beauty of the biotic community. It is wrong if it tends otherwise" (p. 189). When applied to the good life, this definition of life referring to the biosphere might be the most controversial (i.e., unfit for the good life) to ancient philosophers and those who share their views that characterize nonhumans as lesser beings, but the contextual and relational concerns of present scholars suggest that it might still be suitable to the present inquiry. Nonetheless, it seems that we could assess how the lives of a squirrel or group of squirrels are going, for instance, regardless of whether we relate it to human life assessments. It is difficult to say if there is a single concept that pervades these notions or whether four separate terms would better distinguish separate entities, but this questioning provided potential details about the good life to consider.

3.4.2 *Meaning and Sustainability*

Whether any of these notions of life or the descriptions of the good life based on them are compatible with happiness or sustainability is open for debate. However, as demonstrated with counterarguments to well-being theories, it is not clear that any of these thought experiments will necessarily rule out any of the meanings proposed. They might only give some suggestions of exceptions to contemplate or ways to shape our thoughts. It may be the case that each of these meanings is suitable to some extent or might make some contribution to the good life.

3.4.2.1 A Good Life as Vitality

In consideration of vitality, a long life (itself) may make some happy but not everyone. A long life full of happiness is generally appreciated, but some studies contend that peaks and valleys may produce a better life or that happiness at particular times, such as at important life milestones or just before death (i.e., dying happy), may have it best. It is a matter of debate. It may not be possible and perhaps would not be desirable to have a long life of constant happiness (in the sense of being in a perpetual state of thrill, beyond merely contentment, joy, or optimism).

In terms of sustainability, the *Human Development Index* considers longevity (i.e., life expectancy) within its metrics (United Nations Development Programme, 2020). However, having a long life may become a burden to others (Hardwig, 1997). Having children is not always a positive experience or outcome, and pregnancy shaming or shaming individuals who choose to live child-free suggest that there may not be a universal requirement to reproduce or when to stop reproducing as part of a good life. Even if we rule out immortality, longevity could possibly be unsustainable, especially while considering the impact of reproduction, since an increasing population could lead to using up finite resources. So, this life might become self-defeating. For this reason, whether to have children (or how many) has become an important ethical debate in sustainability studies (see more in Hedberg, 2020).

3.4.2.2 A Good Life as a Good Lifestyle

In terms of a good life based on lifestyle, if interpreted as suggesting that the good life is an ethical life, counterexamples can be imagined of unethical activities that might bring happiness to the agent or ethical activities that bring the opposite of happiness, such as taking a joyride in a stolen sportscar or going out of your way to help someone when you are already late to a celebration. These examples suggest that there may be something more to the good life in this sense than mere happiness, even if some ethical actions will also bring happiness.

Whether a lifestyle is sustainable depends on its ecofootprint, its financial budget, and its impact on neighbors, according to the triple bottom line perspective. This sense of the good life is probably the most compatible with sustainable living philosophies regarding individual choices. To the extent that lifestyle choices have more negative eco-impacts, break the budget, or disrupt the neighborhood, they would likely be unsustainable and not represent the good life.

3.4.2.3 A Good Life as (or Including) Well-Being

Debating whether the good life includes or is synonymous with well-being could be an inquiry into whether happiness is an essential component of the good life. It is likely to be a component to some extent and probably not required to be a continuous component, as previously explained as constant happiness. Whether well-being is sustainable is a difficult question to answer without repeating what has been said about happiness or lifestyle. To determine whether well-being is sustainable from a triple bottom line perspective, well-being would then have to include the social and environmental contexts that were debated earlier and also a financial consideration. A life could be sustained without demonstrating well-being either due to those situations described as a life not worth living or if unethical behavior is disqualifying it. A life could be going well without being sustainable to the extent that finite resources are being consumed for success in the moment, but that benefit cannot last indefinitely. Perhaps a sustainable lifestyle might not be able to continue indefinitely (due to mortality), but the intergenerational aspect of sustainability suggests that a sustainable lifestyle would be one that could persist across generations (i.e., would be the sort that people of other generations would imitate).

3.4.2.4 A Good Life in Terms of the (A Good) Biosphere

In terms of the biosphere, not all living beings can experience happiness (probably). So, this meaning of the good life might have to take an ecological connotation to be a good life, such as a thriving ecosystem in some sense. In that way, it is the same as or relates to environmental sustainability. Whether the biosphere gives us happiness is circumstantial. If we limit the notion to human communities, a good life of this type would generally resemble the values of social sustainability, such as equity, inclusion, and respect. Communities might give us happiness at times and not at others. There is some plausibility to describing the good life in each of these ways.

3.4.3 *The Value of Life*

Rather than looking at some *aspects about life* being good (in a similar way to how well-being was evaluated in a previous section), life itself may be valued and for that reason might provide insights into the good life.

Multiple authors present arguments to support valuing life. Though there are other theories that share its name, *vitalism* in the sense I use here is associated with scholars who argue that life has the highest value, such that in any dispute between life and some other thing, life would be chosen. Life's value is sometimes referred to as a *sacredness* or *reverence*. Daniel Callahan and Albert Schweitzer provide some of the most popular arguments here.

3.4.3.1 Religious Vitalism

Callahan (1969) describes at least three arguments for the value of life, presented from religious, secular, and autonomy-based perspectives. Callahan begins with the Judeo-Christian description of the sacredness of life. As life is considered a gift from God, it is sinful to take another's (another person's) life. This principle is most prominently communicated through one of the Ten Commandments—"Thou shall not kill." This vitalistic tenet is still commonly used today to support official positions of the Catholic Church. For instance, a recent papal encyclical used a vitalistic argument to encourage taking action against climate change (Pope Francis, 2015). In counterargument to this reasoning, Callahan describes exceptional instances of acceptable lethality, such as punishment or retribution in certain cases. Further, when someone does not subscribe to these religions, holds the faith without complying with this particular teaching, or in special circumstances, such as self-defense, in war against an aggressor, some instances of suicide, or possibly when persecuted or in sacrifice to the faith, life may not be supported.

3.4.3.2 Secular Vitalism

Besides religious vitalism, Callahan (1969) also reviews Edward Shils' secular reasons to value life and still calls it *sacred*. Callahan suggests that life's sacredness is generated by "experiencing the elemental sensation of vitality and the elemental fear of its extinction" (p. 191). However, not everyone experiences the sensations or fears Shils describes or appreciates those experiences in the way intended. Callahan also cautions that such experiences do not automatically entail moral obligations to protect anyone's life. While this view does not require adopting any certain religion, it is still restricted to individuals who have such sensations and feel morally compelled by them to protect others' lives. If someone does not have these experiences, it might not be possible to make them experience them (to convince them).

3.4.3.3 Vitalism from Autonomy

The third argument Callahan promotes is a contention loosely based on autonomy (Sullivan, 2003). Callahan argues that philosophically speaking, we have to start with some foundational belief, and there may simply be no further support we can give for that type of starting point beyond merely saying it is a foundational belief. Through this line of thinking, Callahan hopes to avoid an endless regress. This argument is said to be founded in autonomy because the proponent is essentially *choosing* to treat the value of life as a premise that is so basic that it cannot be further derived from any other consideration. This argument suffers from a lack of supporting reasoning to describe why someone ought to choose to hold that the value of life is a foundational belief and also is potentially disarmed if someone chooses a different foundational belief. Furthermore, telling someone that the value of life is a foundational belief is not likely to convince them they do not want to die.

3.4.3.4 Metaphysical & Ethical Vitalism

While Callahan provides a medley of vitalistic arguments, Schweitzer's *reverence for life* is one of the most popular accounts. Schweitzer's life-affirming ethics have won a Nobel Peace Prize (arguing against nuclear war), beatification as a living saint (Goodin, 2007), repeatedly voted as one of the top-5 most admired people in the world (Wittner, 1995), and even the critic, Werner Picht, asserts that Schweitzer "remains the greatest moralist of our twentieth-century civilization" (Cicovacki, 2007, p. 2). Schweitzer's life-affirming ethics are summarized in the principle: "Evil is what annihilates, hampers, or hinders life… Goodness is the saving or helping of life, the enabling of whatever life I can to attain its highest development" (1936, p. 233). Schweitzer also states, "It is good to maintain and to encourage life; it is bad to destroy life or to obstruct it" (Cicovacki, 2007, p. 2). Similar to other German philosophers, such as Nietzsche's glorifying of the *will to power*, Schweitzer sees the *will to live* as a nature force within each person and each organism, to which we must give reverence and align or *harmonize* our lives. Cicovacki sees Schweitzer's philosophy as two principles: "(a) do not destroy life, and (b) enhance life" (Cicovacki, 2007, p. 9). Inspired by Schweitzer, Kenneth Goodpaster's theory (1978) of moral consideration for all living beings is based on the value of life, within environmental ethics, which gives it a potentially fundamental role in sustainability philosophy. Schweitzer's view is criticized for a lack of derivation, similar to Shils' secular view that relies on experiencing vitalistic sensations or Callahan's choice-based view of the value of life as an arbitrarily selected starting point. Like the other views, this perspective is not flawless.

While none of the arguments presented here give uncontroversial support to the value of life, they present some of the best attempts available of describing all life as worth living or worthy of respect. Whether more is needed to have a good life is still questionable. Within them, they might also make the case that any life is a good life, but critics might find that problematic. While even a critic might grant that any life is good, it need not be the case that every life demonstrates what is captured in *the good life*.

3.4.4 *Meaning and Value*

We might also investigate whether the different meanings of life reveal something about the value of life. For instance, we might ask:

1. Is being alive (in and of itself) valuable or good?
2. Is a (specific) lifestyle (in and of itself) valuable or good?
3. Is well-being (in and of itself) valuable or good?
4. Is the community of organisms (in and of itself) valuable or good?

3.4.4.1 What Makes Being Alive Good?

While vitalists have been shown to answer the first question affirmatively, some authors compare existing lives or compare never have existed to coming to life, usually with regard to fetuses or future generations (Broome, 2004, 2015; Herstein, 2013). This type of argument might compare being alive to death, as in cases of euthanasia, assisted suicide, or refusal of life-saving treatment. It might be the case that some lives are more valuable than others, but such a conclusion is controversial to suggest. Life would likely usually be good to each being unless depressed or facing tough times perhaps, but even then, a skeptic might wonder if such an individual is mistaken that being alive is not good or valuable to them. Being alive is instrumentally valuable in order to experience any other goods.

3.4.4.2 What Makes a Lifestyle Good?

One's lifestyle is often good to them; otherwise, they would not likely choose to continue living that way, but addictions, misunderstandings, or external factors beyond their control could still interfere with that assumption. To determine whether a specific lifestyle is good on its own, an ethical lifestyle or a sustainable lifestyle might be considered. Ethicists would likely argue that an ethical lifestyle is intrinsically good or instrumentally good for the happiness it brings to the individual or for the peaceful relations it fosters in the community. A sustainable lifestyle is valued instrumentally for the improved chances of survivability it is intended to facilitate, but for subjects who do not wish for themselves or their species to continue, its value might be questioned. As with suicidal tendencies, such a judgment could be criticized as potentially mistaken.

3.4.4.3 What Makes Well-Being Good?

Because improved well-being is typically a reason to seek a change in lifestyle, it seems that well-being is valuable, though it is unclear whether its value is intrinsic or instrumental. However, Hausman (2015), for one, argues that since a life may be otherwise going well but not be admired as a good life or a valuable one, well-being does not incontrovertibly provide a determination for the value of life or a life worth living. However, Hausman further claims that well-being provides no information about the value of a life, which seems a bit overstated. How much pleasure is experienced, desires or objective goods are achieved, how a life is progressing, and how it relates to others and to the environment are all generally indications of a life going well. It remains debatable whether pains, dissatisfaction, or other impairments which leave an individual worse off and typically making the life less admirable might impact the value of life. So, we might grant Hausman's claim if we adopt an egalitarian-vitalistic stance that held that all lives are equally valuable, since then the well-being of those lives would make no difference to their value. However, Hausman does not identify as a vitalist, but rather believes that well-being is a separate concept. Whether adopting a vitalistic stance or not, we could agree that a life with greater well-being would be more valuable than one with less well-being, even if the value is not of the life itself but grounded in its well-being.

3.4.4.4 What Makes the Biosphere Good?

The biosphere is instrumentally good for enabling life across the planet, and biodiversity provides a variety of ecosystem services that are valuable to many organisms. Biodiversity is also instrumentally good for resilience against challenges. Environmental ethicists have not yet been able to definitively declare that the biosphere or biodiversity is intrinsically valuable, but it may be the case that there is nothing with intrinsic value. The instrumental value of life in these various senses give some reason to believe that it is valuable and some ways in which it might be considered more valuable than others.

3.5 Results

As might be expected, this series of word associations exploded into a tremendously complex conceptualization such that each branch of tangential thought seems to open an additional complex of concepts that are controversial and require much more evaluation than is able to be integrated here. After all, the discussion has seemingly encompassed all of life (or at least all of its good parts). Although in some ways the inquiry is unfinished, another conceptual map can be created to illustrate the key learnings produced from the thought experiment and some suggestions that emerged about the guidance it shows.

The conceptual map (Fig. 3.2) shows a dozen ways of describing a good life that arose from analyzing the four definitions of life that were identified according to three levels of the type of subject that might be evaluated—an individual, a relation between multiple entities (intersubjective), or a collective or group (community). Each of these aspects of a good life can be further manifested through a variety of conditions and measured by a variety of indicators (some potentially more meaningful than others).

For instance, individual vitality can be assessed as longevity or life expectancy, and at the community level, life's goodness in terms of vitality could be determined by whether the human species has a good chance of survival. At the intersubjective or relational level, vitality could be measured in consensual reproduction or in a low (or zero) murder rate, either of which can give a sense of life going well in this dimension.

Levels of Good Living Across Various Definitions of Life

	Individual	Intersubjective	Community
Vitality	Longevity	Love	Species (Human) Survivability
Lifestyle	Success or Ethical Action or Lifeplan (Good Citizenship)	Neighborliness or Cooperation	Democracy or Peace
Well-Being	Happiness, Satisfaction, Objective Goods	Happiness for Others (Empathy), Shared Success, Meeting Needs	Mutual Thriving (Social Sustainability) or Justice
Biosphere	Positive Eco-Footprint (Handprint)	Interspecies Cooperation or Mutual Flourishing	Biodiversity or Interspecies Justice (Environmental Sustainability)

Fig. 3.2 This conceptual map depicts four definitions of life in rows and three levels at which a subject might be evaluated in columns. The resulting points of intersection reveal 12 ways that (a) life might be assessed as going well. For example, life's goodness at the individual and community levels might be assessed via longevity and humanity's chance of survival. As another example, intersubjective or relational assessments of life's goodness can be described as cooperative behavior or sharing success with others in the lifestyle and well-being dimensions of life

Since reproduction was criticized earlier as a possibly inappropriate metric, we can instead (or also) think of intersubjective vitality in terms of sharing life with someone else or affirming the life of another as in *love*. Love is manifested in shared vitality, when individuals share life with another so much that they sometimes identify as being *one*, and it can even lead to the procreation sense of increasing vitality, meaning creating more lives. It also seems that a life with more love would be a better life. While that type of love may not be the same relation expressed by not murdering someone, there is a sense that applies even in that case, since noticing or appreciating the shared humanity in another person is a type of love.

In terms of the lifestyle dimension, a good life could be measured through success or ethical actions of individuals, levels of cooperation or neighborliness, or democratic participation or peace at a community level. Each give some sense that life is going well. Well-being can be a subjective, personal measure, and it can refer to how others are recognized or given care in a relational sense. Community well-being can be a measure of life's goodness demonstrated through mutual thriving, justice, and equity, which all relate to social sustainability. To measure life's goodness through the biosphere dimension, the community level is probably most easily identified as environmental sustainability. With interspecies justice and abundant biodiversity, we would be relating well with the biosphere and maybe even with nonliving abiotic entities as well. This connection can also be expressed relationally, rather than collectively, as interspecies cooperation or mutual flourishing. Life's goodness in terms of the biosphere at the individual level returns the discussion to ecofootprint. While ecofootprint is a measure of negative impacts, a *handprint* refers to the positive impacts of one's life on nature (or others). So, it would seem that life is going well if it is good in more of these 12 dimensions and less well if it is bad in any of the dimensions.

3.6 Discussion

Seeking the good life has revealed a few ways to value life and how that might contribute to life's goodness. Whether life is simply good in itself as a foundational belief, perhaps due to its instrumental value for being fundamental to experiencing any other good, or whether it is a gift from God or coming from a fear of death, life is thereby sacred and worthy of reverence to vitalists. However, each of these arguments has a potential weakness that might not convince all critics. That is, they each seem to still require a leap of faith in their current formulations to proclaim life is intrinsically good, but instrumental value might be a sufficient basis.

If someone does not already value ethics, then it is difficult to convince them that it must be considered as a component of well-being that is defined merely by success, happiness, or some other metric, since it is conceptually possible to define these terms with distinct parameters. However, that concession does not fully rule out ethics as part of a good life, if well-being is merely one component of a good life and if a good lifestyle means that a life requires ethical actions or relations with others.

As far as someone seeks the definition of the good life, not just for knowledge's sake but to enact it themselves and to promote that life in others, it is an endeavor of ethics. As far as community and ecological concerns provide important context to or constitution of a good life, some degree of sustainability is manifested or within consideration. So, whether the good life itself includes acting ethically, the pursuit of it is a matter of ethics.

In Schweitzer's principle encouraging affirmation of life as if it were the very purpose of life, there is a sense that enacting sustainability and increasing one's own or others' well-being both are ways to promote life, even if there are other senses of these terms that do not have that meaning. This affirmation might also refer to something like the sense of well-being as life satisfaction—that one should hold "a favorable orientation toward one's life" (Haybron, 2020). Looking at life as a source of goodness or orienting our lives to promote life within us and others seems to have some feasibility of representing a good life. In this way, looking within the concept of life has given some sense of better understanding of the good life. If we orient ourselves toward the 12 conceptualizations of good living, we ought to be getting closer to the good life, to living well. Perhaps peculiarly, when thinking of the good life in this way, the term *good life* may be a redundancy, like *tuna fish* or *free gift*, only if we ignore the bad sides of life, but unpacking the meaning within this alleged pleonasm has revealed insights into our existential purpose that might get us closer to experiencing happiness and progress us toward sustainability.

There still may be more to say if the reader is still asking, "Well, what do I have to *do* to lead a good life?" It seems that the *secret* to living a happy, sustainable, and ethical life is at least partially to *affirm life*. If the 12 aspects of good living are values by which principles can be derived, then perhaps, some corresponding principles of good living could be designed here to create a practical guide for better living. I can only give a provisional or experimental list of suggestions here:

- Act to affirm vitality
  - Act to increase your lifespan
  - Act to respect others' vitality, act with love
  - Act to improve the species' survivability
- Act for a good lifestyle
  - Act ethically
  - Act cooperatively
  - Act peacefully and to encourage peace
- Act to improve well-being in yourself and others
- Act to improve good relations with the natural world and nonhumans to help each other flourish.

Some of these principles might still be too vague to be of much use. For instance, acting ethically requires study of the ethical theories, such as virtue ethics, deontology, consequentialism, or others. The well-being and biosphere dimensions seem to be rather repetitive at the various levels. So, I did not further identify specific principles beyond the general type shown. However, the phrase *act to improve (or achieve)...* might be added in front of the more particular phrases in the diagram to mean something more distinct. *Act to* might be replaced with l*ive to*, but then, the resulting advice that is essentially saying *live to live* becomes questionable without the greater context of this discussion.

As with most of this chapter, each of these principles would need to be tested to find a more suitable wording and further refine the prescriptions, but that task is beyond what I can do here. For instance, acting to increase your lifespan has already been questioned as a potentially circumstantial recommendation than an absolute rule. Moreover, there are actions that might satisfy some of these principles while violating others of them; so, these hypothetical principles are not entirely internally compatible, at least not in the preliminary form I speculate here. Future projects can be devoted to further developing the ideas presented here, giving each deeper consideration, but I wanted to demonstrate a rudimentary way that the thought experiment can be transformed into a practical guide. This conclusion may be disappointing to some readers who were hoping for more concrete advice, but I would want to further scrutinize all of the foundational premises from which these principles are derived before recommending anything further at this time. My goal was to get a better sense of the theory behind such principles; so, hopefully some readers will find that much rewarding, even if like other philosophical works, my chapter ends where it seems I should have begun or ends just as it gets interesting.

3.7 Conclusion

Exploring the definition and value of life in a way that was to be intentionally mindful of the philosophy of well-being and happiness as well as compatible with the concept of sustainability has produced a dozen different potential meanings of *the good life*. I identified a variety of ways to define well-being, happiness, sustainability, and life. Well-being and happiness might be synonyms if only considered as pleasure, but different forms of satisfaction can also distinguish the two. Well-being might also require ethical conduct, and there is at least some chance that it is more important than other factors. If we narrow well-being to acting ethically, it might mean promoting life. Similarly, it might be the case that all actions aiming toward sustainability might ultimately also be promoting life. So, if we seek happiness, well-being, or sustainability, we should determine what it means to promote or affirm life. Affirming life might be necessary to live a good life. If affirming life means to value it, critics may have reason to object to vitalistic arguments of life, but life's instrumental value was not something I could refute at this time. Scrutinizing these conceptions of life and how it might be considered good produced at least 12 ways to assess life to determine whether it is going well. These aspects give a reader the ability to evaluate their life or others' and thereby a variety of ways to seek to improve those lives or the relations between them.

Whether all of 12 aspects must manifest in order to obtain a good life is still up for debate. However, the groundwork here unites some of the foundational conceptualizations of well-being and happiness that tend to be exclusive to individual humans with more communal notions that might better accommodate policy-making pursuits via social and environmental contexts. So, if all of the 12 meanings of a good life must actually manifest in order to have such a life, then policy-making can help to facilitate it at those macro-levels, while individuals work on their contributions at the individual (micro) and intersubjective (meso) levels. This triple level scheme parallels the acknowledgement that greater sustainability requires individual, business, and governmental initiatives to be realized.

To begin to provide more practical prescriptions, 10 tentative principles were derived from the concepts for good living. While they are not necessarily refined into definite rules for a good life, a theoretical bridge from the foundational concepts to an action guide has been created that upon further scrutiny might be developed into something more substantial. So, this chapter provided readers with some philosophy about the good life, in a way that interrelates many concepts and indicators, that might help to improve their decisions when trying to choose to live in a better way.

References

Ackrill, J. L. (1974). Aristotle on eudaimonia. In A. O. Rorty (Ed.), *Essays on Aristotle's ethics* (pp. 15–34). University of California Press.

Ang, J. M. S., Bahlul, R., Betz, M., Bock, G. L., Cowley, C., Dew, R., Elder, A., Gay, W. C., Ketcham, C., Man-to, T., & Lewis, C. D. (Eds.). (2016). *The philosophy of forgiveness* (Vol. 1). Vernon Press.

Athanassoulis, N. (n.d.). Virtue ethics. *Internet encyclopedia of philosophy.* https://iep.utm.edu/virtue/#SH3a

Bethem, J. (2018). Life within energy policy. *Relations: Beyond Anthropocentrism, 6*(1), 69–85. https://www.ledonline.it/index.php/Relations/issue/view/91

Broome, J. (2004). *Weighing lives.*

Broome, J. (2015, February 23). Climate change ethics: Life and death [Presentation]. *Stanford Energy Seminar*, Stanford Precourt Institute of Energy and Stanford Woods Institute for the Environment. https://www.youtube.com/watch?v=gJlkXzYkT3g

Callahan, D. (1969). The sanctity of life. In D. R. Cutler (Ed.), *Updating life and death* (pp. 181–223). Beacon Press.

Cicovacki, P. (2007). Reverence for life: A moral value or the moral value? *Lyceum, 9*(1), 1–10.

Crisp, R. (2017). Well-being. In E. N. Zalta (Ed.), *Stanford encyclopedia of philosophy.* Stanford University. https://plato.stanford.edu/entries/well-being/

Diener, E., Emmons, R. A., Larsen, R. J., & Griffin, S. (1985). The satisfaction with life scale. *Journal of Personality Assessment, 49*, 71–75.

Elkington, J. (1998). *Cannibals with forks: The triple bottom line of 21st century business.* New Society Publishers.

Goodin, D. K. (2007). Schweitzer reconsidered: The applicability of reverence for life as environmental philosophy. *Environmental Ethics, 29*, 403–421.

Goodpaster, K. (1978). On being morally considerable. *The Journal of Philosophy, 75*(6), 308–325.
Hardwig, J. (1997). Is there a duty to die? *The Hastings Center Report, 27*(2), 34–42.
Hausman, D. (2015). *Valuing health: Well-being, freedom, & suffering.*
Haybron, D. (2020). Happiness. In E. N. Zalta (Ed.), *Stanford encyclopedia of philosophy.* Stanford University. https://plato.stanford.edu/entries/happiness/
Hedberg, T. (2020). *The environmental impact of overpopulation: The ethics of procreation.* Routledge.
Herstein, O. J. (2013). Why 'nonexistent people' do not have zero wellbeing but no wellbeing at all. *Journal of Applied Philosophy, 30*(2), 136–145.
Jamieson, D. (1998). Sustainability and beyond. *Ecological Economics, 24*, 183–192.
Kienstra, N., Karskens, M., & Imants, J. (2014). Three approaches to doing philosophy: A proposal for grouping philosophical exercises in classroom teaching. *Metaphilosophy, 45*(2), 288–318.
Kjell, O. N. (2011). Sustainable well-being: A potential synergy between sustainability and well-being research. *Review of General Psychology, 15*(3), 255–266. https://doi.org/10.1037/a0024603
Lee, S. J., & Kim, Y. (2016). Structure of well-being: An exploratory study of the distinction between individual well-being and community well-being and the importance of intersubjective community well-being. In Y. Kee, S. J. Lee, & R. Phillips (Eds.), *Social factors and community well-being* (pp. 13–37). Springer.
Leopold, A. (1949). *A Sand County almanac.* Ballantine Books.
Little, B. R. (2014). Well-doing: Personal projects and the quality of lives. *Theory and Research in Education, 12*(3), 329–346. https://doi.org/10.1177/1477878514545847
Lyubomirsky, S., King, L., & Diener, E. (2005). The benefits of frequent positive affect: Does happiness lead to success? *Psychological Bulletin, 131*(6), 803–855.
Nagel, T. (1972). Aristotle on eudaimonia. *Phronesis, 17*(3), 252–259.
Nozick, R. (1974). *Anarchy, state, and utopia.* Basic Books.
O'Brien, C. (2013). Happiness and sustainability together at last! Sustainable happiness. *Canadian Journal of Education, 36*(4), 228–256.
Parfit, D. (1984). *Reasons and persons.*
Pope Francis. (2015). *Encyclical letter Laudato Si' of the Holy Father Francis on care for our common home.* The Holy See, Libreria Editrice Vaticana. http://w2.vatican.va/content/francesco/en/encyclicals/documents/papa-francesco_20150524_enciclica-laudato-si.html
Schweitzer, A. (1936). The ethics of reverence for life. *Christendom, 1*, 225–239.
Sullivan, M. (2003). The new subjective medicine: Taking the patient's point of view on health care and health. *Social Science & Medicine, 56*, 1595–1604.
United Nations World Commission on Environment and Development. (1987). *Our common future.* http://www.un-documents.net/our-common-future.pdf
United Nations Development Programme. (2020). *Human Development Index (HDI).* http://hdr.undp.org/en/content/human-development-index-hdi
Wittgenstein, L. V. (1953). *Philosophical investigations* (G.E.M. Anscombe, Trans. & R. Rhees, Eds.). Blackwell.
Wittner, L. S. (1995). Blacklisting Schweitzer. *The Bulletin of the Atomic Scientists, 1995* (May/June), 55–61.

Chapter 4
Buddhist Sustainable Development: Inner Happiness as a Direction for Sustainable Development

Sauwalak Kittiprapas

Abstract Sustainable development has largely been a process focused on external interventions. This chapter explores linking happiness and sustainable development via a new concept "Buddhist Sustainable Development" (BSD), suggesting that sustainable development can be achieved through the process of generating inner happiness. The chapter also discusses BSD in relation to other Western approaches, while highlighting happiness classifications in Buddhism and meanings of happiness, human development and real quality of life. The study provides insight on how inner happiness can be a key contributor to a shift towards sustainability. Ultimately, BSD provides alternative approaches to development from conventional approaches, while striving for higher levels of happiness experienced through inner wisdom.

Keywords Buddhist sustainable development · Buddhist economics · Happiness · Inner happiness · Sustainable development · Sustainability

4.1 Introduction

Happiness has been an interesting topic of research, particularly in the context of moving beyond GDP or economic-centric development approaches. The growing interest in happiness research has been supported by an increasing numbers of studies and discussions. Yet, the link between happiness and sustainable development, in particular, is not clearly evident. With a global commitment to achieving the Sustainable Development Goals (SDGs) by 2030, sustainable development is a serious movement. This study argues happiness development should support or lead sustainable development efforts. However, happiness has a weak presence or is missing within conventional sustainable development approaches (Kittiprapas, 2015a). For instance, a significant research gap exists in that "neither mental well-being nor the spiritual has been taken into consideration by the SDGs through their

S. Kittiprapas (✉)
International Research Associate for Happy Societies (IRAH) and Faculty of Economics, Rangsit University, Lak Hok, Thailand
e-mail: sauwalak.k@rsu.ac.th

S. Cloutier et al. (eds.), *Linking Sustainability and Happiness*, Community Quality-of-Life and Well-Being,
https://doi.org/10.1007/978-3-030-89559-4_4

17 goals… there is a big gap between subjective and objective realities of sustainable development" (Sangasumana, 2019, p. 247). Sangosama also points out that, "even though many indices have been formulated for the purpose of measuring happiness, the actual measure of relevant subjective wellbeing has not been taken into consideration" (p. 247).

This chapter aims to close the research gap by introducing a new concept linking subjective inner happiness and sustainable development: Buddhist Sustainable Development (BSD). BSD is grounded in a Buddhist approach to development, highlighting the inner change needed to achieve higher levels of happiness necessary for sustainability. BSD is different from conventionally used sustainable development frameworks in that it focuses on human behaviors resulting in sustainability challenges, rather than ignoring human desire as the root cause of unsustainability.

The chapter is organized in five parts: (1) a background on BSD as a concept, (2) an exploration of happiness explained as different levels from a Buddhist perspective and their relation to BSD, (3) a comparison of BSD to other development frameworks that include happiness, (4) the implications of BSD for sustainable development, and (5) concluding remarks. The intention of this chapter is to introduce the new BSD framework, with an explanation of related concepts, as an alternative to mainstream sustainable development and to show how this new sustainable development concept links with different kinds of happiness. It is also suggested that this concept, by leading to different outcomes than development approaches that have created problems, will be useful in contributing to more discussions from diverse perspectives in the fields of sustainability and happiness.

4.2 Background and Explanation of the BSD Concept

To my knowledge, a novel idea mentioning the possible link between *sustainable happiness*[1] and sustainable development was found in Ven. Payutto' s works (1995, 2006). The works question Brundtland's definition of sustainable development and the ineffectiveness of associated global efforts to achieve sustainable development goals, given a lack of real or sustainable happiness. He discusses the conceptual contradiction of compromising sustainable development for future generations. Specifically, current generations are unable to compromise their (temporary) happiness for a sustainable future (Payutto, 2006). Thus, sustainable (or real) happiness in Buddhist terms is crucial for sustainable development (Payutto, 1995). Inspired by the idea and related works, the BSD framework was formulated as a part of my research project, entitled *Buddhist Sustainable Development: Inner Happiness and World Peace*, which was submitted as a proposal to the World Buddhist University, and was funded and titled in early 2013. During the research process, the BSD

[1] This term *sustainable happiness*, based on Buddhist perspective, is defined as happiness from inside (inner liberalization) with no suffering or the opposite to unsustainable happiness which is temporary.

concept papers were being discussed and exchanged in related conferences among international Buddhist scholars in Asian countries and Thailand (including a forum for academics and development researchers/organizations on the UN International Happiness Day in 2015). This chapter builds on those works as well as existing happiness and sustainable development concepts discussed within academic research, by integrating inner happiness teachings from Buddhist monks' works and practices and their link to sustainable development, which has become a global agenda.

Generally, happiness and sustainable outcomes are often treated as separate. However, there may be a causal relation between sustainability and human happiness, as found in recent publications (Berejnoi et al., 2020; Cloutier & Pfeiffer, 2015; Kittiprapas, 2016, 2018a; O'Brien, 2008; Payutto, 1995; Ulluwishewa, 2016; Zidansek, 2007) in response to ineffective sustainable development methods so far. In that context, BSD suggests internal human transformation as a key to a sustainable future, given that there have been ineffective sustainability outcomes across nearly four decades. It suggests inner happiness as a missing link within many current sustainable development frameworks, including the SDGs, and is a pathway to achieving sustainable development (Kittiprapas, 2015a). In other words, BSD is conceptualized as human-centered sustainable development, driven by inner happiness (Kittiprapas, 2018a). As the human mind is neglected in key pillars of sustainable development (which focuses on economic, social, environment, and recently the SDSN (2013) adding on the fourth pillar of good governance), BSD focuses on the inner-human dimensions or internal human transformation as a critical factor for sustainable development and a starting point for change.

After the description of happiness in Buddhism and BSD, I will provide some more details and discussion around related concepts to identify similarities and differences amongst approaches. The intention is to provide some information about gaps and complementary aspects. Hopefully, BSD, together with emerging similar ideas, contributes to existing discussions and movements toward achieving global sustainable development goals.

4.3 Happiness in Buddhism and Buddhist Sustainable Development Concept

Kittipprapas (2016, 2018a) classified three schools of thought on sustainable development: economic-based (environmental economics and green economy); ecological-based (ecological economics and deep ecology); and Buddhist-based concepts (Buddhist Sustainable Development, BSD; Gross National Happiness, GNH; and Sufficiency Economy Philosophy, SEP). While economics and ecological concepts both deal with physical resources for sustainable development (but in different approaches and views), Buddhist-based concepts focus more on the human mind and behavior. However, the Buddhist-based concept itself has some differences; for example, GNH focuses on happiness as the overall goal (while environmental preservation is one of four pillars to achieve it), while BSD sets sustainable development as the end goal with the process of internal happiness transformation as the means. The BSD, in particular, deals with the human mind as the root cause of unsustainable development.

Happiness in Buddhism consists of different types and ranks; an individual is encouraged to move from the lowest rank of happiness to the highest one, which is considered as development. Although levels of happiness can be classified into many detailed ranks, next, I classify happiness into three levels and explain their principles, based on Payutto (2011) and Kittiprapas (2015b, 2015c): (1) sensual or physical, (2) mind, and (3) wisdom or inner liberation.

Level 1. Happiness at the Sensual Pleasure or Physical Level

Happiness derived from sensual pleasures or on the physical level is considered as the lowest level of happiness, obtained from the five senses: eye, ear, nose, tongue, and body. Such forms include pleasant noises, sight, good smells, tastes, and comfortable touches. Pleasure is experienced from enjoying these things on the physical level. People have desires to seek such pleasures and avoid pain. This includes happiness from material acquisition and fulfilled wants (beyond basic needs). It is the lowest level of happiness and unsustainable in its nature as it depends on many factors external to the human being for which one has to search. Desires to meet this type of happiness can result in suffering when desires are not met; or even when the desires are met, happiness can be temporary and fade away over time (which are also explained by theories of social comparison/aspiration and adaptation). This level of happiness is not considered real happiness in Buddhism, so people have to be trained to move up to level 2 and 3.

Level 2. Happiness at the Mind Level

Level 2 happiness can be generated by the mind, not requiring any outside objects. Therefore, this level of happiness is considered as higher and more sustainable than the physical level which depends on other factors outside oneself. With mind-based happiness, one can be happy from a mind containing virtues, peace and fresh perspectives.

Instead of being happy from acquisitions, individuals can be happy from giving away, sharing and acquiring less. Therefore, people can be generous toward others rather than selfish or self-centered. Both individual happiness (from mind development) as well as societal happiness increase, as the givers and receivers are both happy.

Mind-based happiness is a higher level than the first type of happiness because it is driven by a higher quality or good desires and a mind developed or trained to be so, not driven by defilements or personal benefit only. This type of desire is called good will/aspiration. People can experience desires to be good, to be knowledgeable, to be creative, to be useful, to give or share what they have with others, to help people out of suffering, to do good things or good work (rather than being concerned only with monetary returns). People can be happy from a calm mind, natural appreciation, positive thinking, being generous/helping others, working with good aspiration, showing compassion and loving kindness, and less attachment to oneself. Such outcomes could benefit societal development and happiness.

Mind development practices can bring in a higher quality of mind characterized by power, clarity, and peace. These are conditions for the establishment of various virtues that can be used for reflection, investigation, and contemplation through wisdom (Payutto, 2011).

Level 3. Happiness at the Wisdom or Inner Liberation Level

Happiness at this level is characterized by wisdom or insight, with full understanding of natural law from the three characteristics of truth[2]: impermanence (natural changes), the state of suffering, and non-existing (or selflessness). Realizing that these characteristics are not worthy to attach to, human beings then have purified minds free from any attachment. Therefore, they are completely free from suffering as they have insight to understand all natural laws; namely, natural changes as well as the relationship of cause and effect to neutrally examine things as they are, without additional emotion but with right mindfulness. They have right views that: human beings are part of interconnected natural systems of cause and effect; all beings (human and non-human) are friends under the same natural laws; and, most importantly, happiness with liberation from the inside brings in the best and perfect life (Payutto, 2013).

This type of high quality of happiness is from inner liberation, not from materials or dependence on outside factors. This is the highest level of happiness and more sustainable than the two lower levels of happiness. At the highest point of this happiness level, nothing can cause suffering to a person whose wisdom is able to solve all problems wisely with the mind of emptiness and detachment from all defilements. The individual who has progressed to reach this highest level is completely awakened or enlightened, with natural freshness and joyfulness (Payutto, 2013, p.13). Those who enjoy this highest level of happiness would not return to attach to lower levels of happiness. They will neutrally understand everything under the law of nature without suffering; thus, this type of happiness is a sustainable form that human beings can create from internalization.

If individual happiness is supervised by wisdom, the search for material things can be utilized positively for the benefit of others and society. This behavior can bring in more peace and happiness to society. Therefore, individual and societal happiness are not in conflict in this case while the first type of individual happiness with self-centered goals may cause societal unhappiness. That is why human beings have to be trained in moving up to the higher level of happiness[3]; thus, happiness in Buddhism is not static but dynamic aiming at reaching the highest goal. The three levels of happiness mentioned above are briefly covered in Table 4.1, and more implications for sustainable development are discussed in Sect. 4.4.

---

[2] This is the three principles in Dhamma teaching in Buddhism.

[3] Four foundations of mindfulness is a guide to progress internal development of human beings to reach higher levels of happiness until the highest point of sustainable happiness.

Table 4.1 Characteristics and outcomes for three levels of happiness and sustainability

Levels of happiness	Characteristics	Outcome
3rd level—Wisdom-based happiness	Happiness from understanding natural truths, conditionality and causal relationship, un-attached to suffering leading to real or sustainable happiness. Understanding human-nature interdependence, treating nature with care and compassion. Balanced living and relationship with nature Material beyond basic needs are used for benefit others and to preserve nature. Wise consumption and moderation benefits all beings	Peace and sustainable happiness, at individual and societal levels. Happiness from least resources used. People experience selflessness so they would do no harm to others and nature. Supervised by wisdom, materials are utilized positively for society. Development is directed by right views. Sustainability can be achieved
2nd level—Mind-based happiness	Happiness from good mind and willingness to help others with loving kindness, empathy/compassion, sharing and giving (less acquisition and more donation). Mind is calm, fresh and joyful. Although this is less self-centered than the first level, this type of happiness may involve some expectation of good results or attachment that can create suffering for individuals (though less than the first level)	More sharing within society and less conflict than the 1st level. Less resources exploitation for individual needs. More useful activities and work for the public at large, leading to sharing and caring societies. Resources can be better distributed to those in need
1st level—Physical-based happiness	Happiness from sensual pleasures and material acquisition (i.e. for wealth, income, private goods, etc.). Increasing wants beyond basic needs will create problems and unhappiness. High competition and self-centeredness can cause conflicts with others and nature. This is temporary happiness, depending on external factors. It is not real happiness as it involves excessive suffering for individual and others, and can destroy nature for greed	Unsustainable happiness, stress, conflict, and tension to obtain possessions driven by unlimited wants for oneself. Individual sufferings, social conflicts, and environmental deterioration. Over consumption and production and excessive use of resources leading to unsustainable development

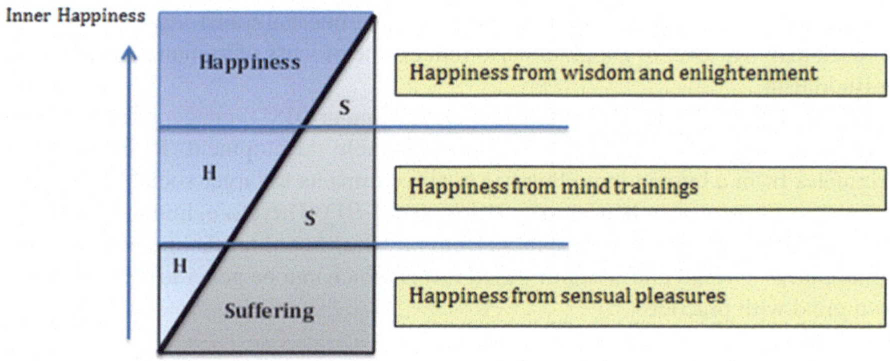

Fig. 4.1 Degree of happiness and suffering at different happiness levels (Kittiprapas, 2018a)

Within the Buddhist happiness concept above, individuals are encouraged to develop their happiness from the lowest level to the highest level: from material acquisition to mind-based happiness and finally to wisdom/insights. The higher levels of happiness are more sustainable with less dependence on outside factors. Lower levels of happiness are unsustainable because they are mixed with suffering (when desires are not met); although, if desires are met, that happiness would be temporary due to effects from adaptation, aspiration or social comparison.

Figure 4.1 seeks to illustrate the happiness concept in Buddhism. It simply shows how happiness in higher levels has increased with decreases in suffering. Suffering and happiness are counter effects in the mind. When there is no suffering, happiness can be occupied in one's mind. Higher inner happiness associates with higher level of overall happiness.

The bottom level is happiness from sensual pleasures, which involves a lot of suffering, so it cannot generate much true happiness. The middle level represents the mind-based level that involves more happiness generated inside by oneself. Suffering drops with the degree of increased inner happiness. Inner happiness increases continuously when moving towards a higher level of happiness. That is why people should aim for higher levels of happiness which is more sustainable. When inner happiness increases to the highest point (enlightenment), there is no room left for suffering. Thus, from Buddhist teaching, this is the direction for individuals' happiness development: to move from the lower level of happiness to the higher in order to achieve more sustainable happiness. This kind of happiness can also lead to effective sustainable development because people can be happy with less wants, less exploitation, and less harm to natural resources; thus, from BSD perspective, this is a pathway toward sustainability.

For many people, happiness development beyond the first level of material dependency is encouraged (even though they may not reach the highest level) as an individual can be developed to change levels/types of happiness. Higher quality happiness can be achieved through non-acquisition, which is mind and wisdom-based happiness (inner happiness). Inner happiness is more important and a more sustainable pattern as it is usually accompanied by less suffering. The internal transformation required to reach a higher level of happiness also reflects the process of human development[4] in Buddhism.

In essence, the Buddhist Sustainable Development (BSD) concept suggests that inner happiness is the key for effective sustainable development. Real happiness originates from a human mind that has positive impacts to larger social and natural system (as illustrated in Kittiprapas, 2018a, p. 90–91). Therefore, human beings are encouraged to develop their happiness from physical-based levels into higher forms of happiness at mind and wisdom-based levels which can be generated inside ones' own mind with practice.

4.4 A Review Discussion of BSD Compared with Other Related Concepts

The profound thought for the real value of life with high qualities of happiness are in traditional wisdom from both the East and the West. Therefore, it is not surprising to see that values are reflected in some similar ideas or philosophies in different parts of the world, though details may be varied. This section discusses some similarities and differences between BSD and other Western concepts, including recent works linking happiness and sustainability, which have gradually emerged in the last decade.

The ancient western concept, Aristotle's eudaimonia approach, also accounts for the spiritual element and the feeling of contentment for happiness and considered happiness as a path for wisdom and morality to attain a good life. Thus, traditional Western and Eastern philosophies seem to have similar ideas on happiness and the goal of life: happiness at a high level of wisdom/morality and a good life, which includes self-sufficiency and ethics (Kittiprapas, 2018a).

Both ancient Western and Buddhist philosophies regarded hedonism and greed as threats to social stability, and both believed that tensions between the individual and the community could be moderated through an ethic of virtue, in which individuals live their lives in accordance with human nature and social realities. In addition, both believed that the *right path* (for Buddhism) and *virtue* (for Aristotelianism) require training, education, practice, and cultivation of the mind (Sachs, 2013, p. 84).

While there are some similarities, the path or approaches to achieve a good life, as reflected in the two philosophies, may not be exactly the same. For example, Buddhism believes in self-practice (through practicing mindfulness/meditation) to reach the highest goal of wisdom or insights until self-enlightenment, which goes beyond the moral-goal level. To move along this right path, ethics or virtues will emerge automatically on the way (with mindfulness) but not as the ultimate goal.[5]

---

[4] This concept of human development is different from the general concept of human development (such as that of UNDP) that only relate to objective well-being but not focus on human internalization as in Buddhist approach.

[5] This is the author's personal interpretation/understanding.

This is beyond the idea of returning to *virtue ethics* as a strategy to promote happiness in the society as noted by Sachs (2013), as the end goal of Buddhism is full wisdom/ insight rather than accompanying virtue ethics that would occur as a result.[6] Buddhist philosophy focuses on wisdom through understanding the ultimate truths of nature as the most important goal and process in achieving absolute happiness. This is the process toward self-realization through insights.

Although ancient Western and Eastern Buddhist concepts have some similarities regarding happiness and the concept of good life, they seem to be different in economic assumptions and systems, in addition to ideas such as human goals and the role of wisdom. Buddhist Economics considers economics as a means to improve real quality of life and human development, not as the end by itself (Payutto, 1992, p. 53); economic activities must support good and the noble life of individuals and societies. Buddhism is focused on human development through wisdom cultivation for liberalization from suffering, which is different from modern human development concepts. Thus, Buddhist economics is a holistic development approach aiming for true well-being (Kittiprapas, 2017).

Another similarity between Aristotle's and the Buddhist concept is the differentiation between natural needs (or real needs) and unnatural wants (which are unlimited), as noted in Puntasen (2001). Regrettably, modern economists have neglected this crucial point that has a definite impacts on human, global, and sustainable development.

In fact, it is crucial to distinguish between real or basic needs necessary for life and unnecessary wants or unlimited greed, which has great impacts on resources. In Buddhist economics (Payutto, 1992), people should have or be taught the insight to know which level is moderation or an appropriate level of consumption. This insight could have great impacts for sustainable development. By not attaching to those beyond basic needs, one can free oneself from material-based happiness and move up to higher levels of happiness, as shown in Kittiprapas (2018a).

Regarding differentiating between a high and a low quality of happiness, Mill's (1863) concept of happiness may be similar to that of Buddhism in the sense that it differentiates levels of happiness and places more weight on a higher quality of happiness. Mill also considers happiness from morals and wisdom (higher forms of pleasure) to be more superior than happiness from the physical level (a lower form of pleasure). People who have chances to experience both levels of happiness would prefer the higher one with higher quality; those who prefer simple or lower pleasures may do so because they have never experienced higher pleasures. Thus, Mill's idea of happiness seems very similar to Buddhist concepts of happiness development.

To compare Buddhism's ladders of happiness development with Maslow's hierarchy of needs (1943) as shown in Fig. 4.2, there are some similarities and differences. They are similar in that they both realize the importance of basic physical needs, such as food and water as indicated in the bottom level of Maslow's, and the top of pyramid relating to wisdom, such as problem solving, lack of prejudice and acceptance of facts (reality).

[6] This is also the author's interpretation/understanding.

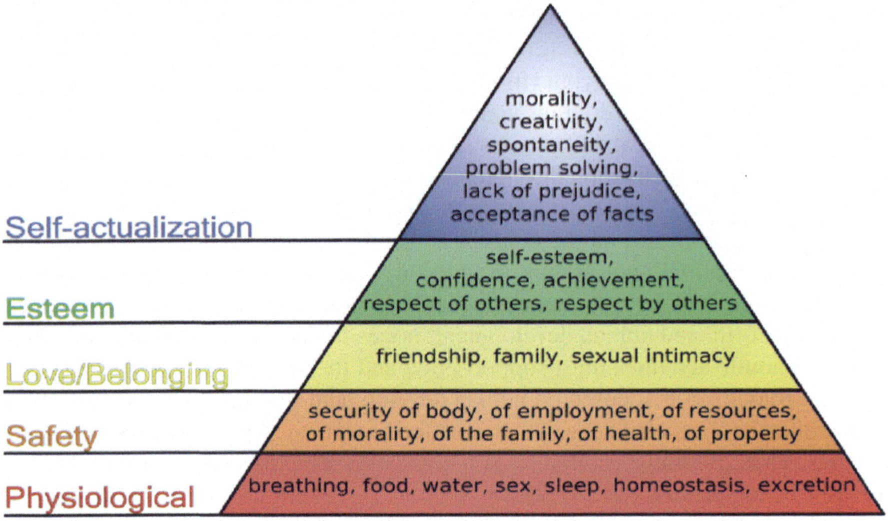

Fig. 4.2 Maslow's hierarchy of human needs. *Source* https://5touche.wordpress.com/2017/09/17/5-stages-of-maslows-hierarchy-of-needs/

However, some details toward human development of the two approaches are different, although the end goal for self-realization seems similar. Through the process of human development in Maslow's theory of human needs, the low level starts with physical needs, the next one is safe surroundings, and then social needs (e.g., social relations, being admired and respected) or psychological needs (love, self-esteem and achievement) and then to moral needs at the top level of the pyramid. Apart from the highest level, self-actualization—the only level at which human beings can create and meet the needs by themselves—the other levels of needs all depend on external factors such as acquisition, surroundings, or social expectations.

Maslow prioritizes the first level of needs as the most important to be fulfilled before being able to develop needs in higher ranks, which also depend on external factors (social needs, love, acceptance, respect, and expectations from others). Buddhism places a greater importance on inner happiness, which is internal or less dependent on external factors and other people that could lead to unsustainable happiness. In fact, some of those beyond the lowest physical level and before reaching Maslow's top level of self-actualization (Fig. 4.2), may be close to the second rank of happiness in BSD (Fig. 4.1), or the mind-based level, which depends less on physical-based happiness and more on giving and, perhaps, still needing acceptance and expectation from others. Both mind-based happiness, from BSD, and the range from just above physical to self-actualization in Maslow's pyramid, will still likely involve suffering and not yet fully sustainable happiness as their expectation may not be met and they do not reach full internal liberation.

In addition, these two approaches may be different in considering quality and quantity. Buddhist happiness may give more weight to quality aspects such as quality of mind (purified, calm, and clear) to obtain wisdom, which are an internal process while external needs beyond basic needs are not necessary. Buddhism places importance to happiness on liberation (from insights) that one can achieve independently, and not on external dependence. Only at the top of the pyramid of self-actualization in Maslow's hierarchy, can humans be less dependent on others to achieve freedom in happiness; that is similar to happiness at the top level in BSD. In sum, a key difference between the two approaches is in the process to reach the top level of self-actualization, in that Buddhism requires internal transformation process to move up the levels of human development.

Consistent with BSD, Ulluwishewa (2016) focusing on spiritual well-being, noted that there is increased recognition in the significance of inner change. He also quoted a number of Western scholars mentioning inner transformation of human being for sustainable development (p.156). Similarly, these spiritual foci indicate human behavior as the root cause to be changed rather than external changes; i.e., technology, institutional, social, economic and political issues. It also points out that a transformation of the relationship of human beings with nature can lay the foundation for sustainability and happiness. That is also similar to Payutto's (2013) suggestion that humans must have right views toward life and nature, suggesting a holistic approach with human spiritual development as a foundation for sustainable happiness for all and for sustainable development (Kittiprapas, 2015a).

Another similarity between Ulluwishewa (2016) and Buddhist happiness concepts is that Ulluwishewa (2016) also separates what is called *matter-self* and *energy-self* for people to transform themselves from matter-self to energy-self through spiritual approaches. Thus, their mind and behavior will change (from self-centeredness to selflessness) for more loving kindness toward others and nature. This seems similar to Buddhist happiness development transformations from the lowest to the highest happiness levels. The paper also mentions energy-self as the highest need in Maslow's need of self-actualization. It also points out the capacity to transform matter-self to energy-self and, as supported by neuroscience, pleasure is not happiness in the sense of seeking pleasure that can lead to unsustainable happiness and unsustainable development.

Similarly, scholars often argue that modern culture, characterized by the consumerism within contemporary capitalism, promotes and values hedonic well-being over the longer-term gains of meaning and purpose. Unsustainable lifestyles can cause environmental and economic crises as well as intra-national and international conflicts. Therefore, sustainability is seen as a domain of collective well-being. As noted in Puntasen (2018), since the age of industrial revolution in the eighteenth century, material wealth was interpreted as the main source of hedonic happiness and in modern economics it has become increasingly synonymous with happiness.

Similar to Ulluwishewa (2016), Berejnoi et al. (2020) also noted that sustainability science has given insufficient attention to the role of spiritual well-being in sustainability. The study examined students' sustainability behavior using sustainability classroom curriculum, including contemplative practices (CPs), finding increased traits indicative of spiritual development and well-being and happiness, and that students increase their self-compassion and happiness. Spiritual growth cultivated through contemplative practices (CPs) increased motivation to engage in sustainability behavior though it does not empirically increase subjective well-being.

Consistently, BSD requires mind-based development as a starting point for change. As quoted in Payutto's (1994) speech to UNESCO in the occasion of receiving the prize for Peace Education,

> All such changes must begin in the mind. They can only be achieved by the mind that enjoys inner peace, freedom, and happiness.
>
> By directing our efforts more inward through practices such as skillful reflection and meditation, it is possible to find a subtler and more independent kind of happiness, born from calmness and understanding, which allows us to rise above exploitative actions. Ultimately, human development leads to the full realization of truth and complete understanding of the interdependence of all things: our lives, society and physical environment.

The speech reflects the linkage of internal happiness and external social contexts and environment. Mindfulness[7] is a key method for inner development to reach the highest level of wisdom and happiness in Buddhism (Kittiprapas, 2016, 2018a). While mindfulness has recently gained more recognition in practice, even in Western world, only a few studies have attempted to examine mindfulness and sustainability linkages, as noted in Wamsler et al. (2018). Through extensive literature review, they draw on notions such as *sustainability from within, ecological mindfulness, organizational mindfulness*, and *contemplative practices* that have been neglected in sustainability science and teaching; and mindfulness is becoming a core concept in sustainability science, practice, and teaching. The paper cites Kjell (2011) and Kajikawa (2008) which independently found that sustainability and well-being research are separate, so subjective human aspects tend to be ignored in sustainability research.

Wamsler et al. (2018) mention scientific support for the relationship between mindfulness and human-nature connection, providing a framework that links mindfulness on the individual scale to collective or planetary well-being at a global scale of sustainability. These aspects are culturally shaped and include: (1) subjective well-being; (2) activation of (intrinsic/non-materialistic) core values; (3) consumption and sustainable behavior; (4) the human–nature connection; (5) equity issues; (6) social activism; and (7) deliberate, flexible, and adaptive responses to climate change.

The community level can be a unit for experimentation to explore the happiness and sustainability linkage. While general community approaches highlight the importance of the surrounding environment and some relationship to happiness/subjective well-being, Cloutier and Pfeiffer (2015) argue that conventional sustainable community development approaches neglect happiness as a goal, and the paper attempts to put forward a happiness-centric approach to sustainable development. With the goal to improve community-level happiness, community residents may be easier involved in sustainable community development projects; thus, easier to achieve developmental goals. They developed the Sustainability through Happiness Framework (STHF), an approach that puts community happiness as a goal in communities

[7] Four foundations of mindfulness are pathways to increase inner happiness in practice, as explained in Kittiprapas' (2018a, p.109–114) Annex: How to Increase Inner Happiness.

and a set process of intervention to improve happiness in 5 iterative stages: happiness visioning, public participation, a happiness profit inventory, and systems planning and sustainability interventions Ibid). Communities also shape happiness by indirectly cultivating social relationships and directly through their inclusion of amenities that affect wellbeing. In Cloutier et al. (2019), cultivating community well-being is an approach rooted in the perspective that community wellbeing could benefit not only from improved resources and infrastructure, but also from processes that promote humans willing to build and connect with community. That implies the linkage of community well-being and subjective human well-being.

In addition, there is a growing interest in how pro-environmental behavior affects people's subjective well-being, and in turn, how people's emotional states can be effectively leveraged to nudge them into behaving in more environmentally friendly ways (Krekel & MacKerron, 2020). This also implies the linkage of environment and subjective well-being. Similarly, Zidansek (2007) discusses the linkage between happiness and sustainable development and suggests that individuals should shift their values of happiness from non-sustainable ones to sustainable ones (to post materialistic values). In that value change, people can become happier and the world can be more sustainable simultaneously. He notes that if an individual can find a more fulfilling way to satisfy basic human needs that increases both happiness and sustainability, a step toward a more sustainable society would be made. The gap identified by Zidanek, in shifting happiness and values for sustainability, is one that the BSD inner happiness approach can address.

However, various studies mentioning happiness and sustainability relationship may have different focuses in details with respect to the linkage between subjective well-being and sustainability. Although some community development frameworks have recently tried to extend the concept of happiness for community well-being and sustainability, they seem to show the relationship between the two rather than a focus in individual's inner happiness as the starting point toward sustainability and a process to reach the goal of sustainable development. They seem to set happiness as the goal, rather than the process of internal development as in BSD. Another difference is that some works may define the term sustainable happiness in relation to environment and others, rather than sustainable happiness inside individual mind, as defined by BSD. For example, O'Brien (2008) defined sustainable happiness as the pursuit of happiness that does not exploit other people, the environment, or future generations; while Kittiprapas (2018a) refers to sustainable happiness based on Buddhist concepts, as higher levels of happiness or inner happiness (from mind and wisdom-based). In other words, BSD refers to happiness beyond physical-based temporary happiness that is easily affected by adaptation and aspiration.

Thus, the difference between O'Brien (2008) and BSD is that the former sets happiness as a goal to pursue with a means not to exploit others and environment, while the latter sets sustainable development as a goal to pursue through inner shifts toward being happy with mind and wisdom based happiness (or inner happiness) as a means to achieve that goal. Therefore, while both link happiness and sustainable development, they are slightly different in ends and means as well as definitions of the term sustainable happiness.

Through a review of these literatures, it can be concluded there is a missing link of the inner-human focus of sustainable development in conventional concepts and existing approaches, while there is gradually an increase in recognition of the importance of inner change for sustainability from scholars in different parts of the world. This emergingly reflects a joint realization of potential causal relations to unsustainable development from low levels of happiness. In this stream, BSD explicitly defines inner happiness as a direction for sustainability, as ignoring this point in the conventional approach has led to the ineffectiveness of achieving sustainable development for a long time. Specifically, using BSD means the inclusion of internal human transformation towards higher levels of happiness (inner happiness) to achieve sustainability goals.

4.5 BSD and Implications to Sustainable Development

Effective sustainable development needs right views towards life and nature. As mentioned in Payutto (2012), *Right views* is one of key 8 paths of the Buddha's teaching of the middle path—the guidance to reach the highest wisdom/insight and enlightenment-and right views is the origin of all development which can reflect the quality of real development. Inner happiness, a realization of right views, would result in the modest use of resources and a balanced lifestyle between human and environment. Thus, BSD implies that maximizing human well-being is coupled with low consumption or less use of scarce resources. Kittiprapas (2020) found that inner happiness indicators, such as the feeling of contentment or moderation and suffering management, have significantly positive effects to happiness in a Buddhist community which preserved their environment for a long time. In other words, a relationship between sustainable community development and inner happiness is empirically evident in that case.

The key difference between BSD and other sustainable development approaches is the emphasis on the inner path for sustainable development. It requires human transformation and behavior change. BSD implies that being less concerned for oneself and less dependent on materials with higher inner happiness can lead to more societal happiness with peace and sustainable development. The key implication from this development concept is that people can be happy with less consumption and acquisition; thus, there is no need to accelerate consumption or economic growth or the overwhelming use of resources, which is counter to the sustainable development path. So far, the world has experienced the excessive use of resources at more than optimum levels of happiness due to overconsumption and production from unlimited desires and greed. Therefore, an internal human focus of happiness development is essential for effective sustainable development.

BSD, as proposed in Kittiprapas (2016), set the process of generating internal happiness as a pathway to reaching sustainability goals (although ultimate or real happiness is also a goal by itself). The key element of the concept is using happiness concepts from Buddhism, which come from higher development of the human mind and wisdom with less demand in materials. This process of limited demand by consumers (with higher level of happiness), is missing in the concept of modern economics and development concepts, which believe that higher consumption/materials implies higher wellbeing. In contrast, inner happiness with less consumption and resource use can lead to higher subjective well-being (Kittiprapas, 2015a). The conventional economic and sustainable development frameworks have neglected the focus of human-mind or mental state which is the source of suffering. In particular, this is missed in the widely used global definition of sustainable development of Brundtland Commission, from *Our Common Future* in 1987 (p. 41), that is:

> Sustainable development is development that meets the needs of the present without compromising the ability of future generations to meet their own needs

The above concept considers meeting the needs of present and future generations only without due consideration to unlimited desires of human beings. Payutto (2006) criticizes the Brundtland concept and indicate that inner or human's sustainable happiness is a pre-requisite for achieving the sustainable goal; that Brundtland definition has missed the crucial concept of sustainable happiness and that is why sustainable development has not been effectively achieved so far.

In other words, the Brundtland definition set a high goal but lacked the process of *how to* achieve this development goal (Kittiprapas, 2018a). The internal transformation in happiness in Buddhism is a way or *how to* reach the goal of sustainability. Therefore, the framework for sustainable development should add more human elements in addition to the existing pillars of economic, social, and environmental, because the unsustainable development problem arises from human behavior (Kittiprapas, 2015a, 2016). To consider the human element in BSD would make sustainable development frameworks more holistic.

Buddhist Sustainable Development (BSD), based on Buddhist economics, encourages people to realize real value of products/materials to be consumed only for improving real quality of life, rather than to meet defilements that never end or unlimited wants in modern economics. The BSD concept classifies human desires at the physical level into two types: (1) those really needed for human well-being, and (2) unlimited wants not necessary for real well-being (Kittiprapas, 2018a). When there are internal conflicts in choosing between the two desires with resource limitation, one should go for the first type for improving true human well-being.

Human beings cannot live without the first type of physical basic needs for survival, but can live happily without the second type of materials which are essentially luxuries. As noted in Payutto (1992), contentment implies the absence of artificial wants but desires for true wellbeing remain. Thus, for content people with inner happiness in mind, the second type of wants is not necessary or considered as artificial wants. Limiting desires to only real needs (or close to) for a standard quality of life could reduce the overwhelming consumption in luxury items as one should be mindful to consume only what is really good for life (or real quality of life). Human beings would have fewer wants for physical/materialistic things and less consumption, leading to less production and less use of natural resources. Thus, this concept, based on Buddhism, indicates that we can maximize human well-being or happiness while minimizing consumption or using less scarce resources. Payutto's (1992) Buddhist economics indicates that the equilibrium is the point that human satisfaction meets true well-being (real needs), bringing in the *right amount* of consumption. In other words, the right amount or *appropriate level* or *moderation* can bring in human satisfaction or happiness. This is the right demand for life or *optimal for life* which is the concept of right living. In fact, an increase in over consumption and excessive use of resources may even result in the declining individual happiness (as it would involve increased suffering due to stress, burden, conflicts, unhealthy, etc., as depicted in Kittiprapas, 2018a, pp. 24–25); thus, using resources beyond optimum levels is wasteful or even harmful to happiness.

This concept implies that there is no need for over-spending and over-producing or exploitation of resources, which do not increase real quality of life[8] and happiness. For economic development, this concept may be called *economics of moderation*; that is, to know what is an appropriate level or what is enough for each individual's conditions. As everyone has different life conditions, knowing what level is appropriate or is *enough* in acquisition requires wisdom. For example, people should know appropriate financial spending for households' certain levels of income, having mindful consumption for good health, work-life balance, and appropriate material accumulation as not to be burdened or stressful and harmful to others and nature.

Production at a level higher than satisfying a healthy human life can result in an over-utilization of resources. Eventually, it can also result in threatening the existence of a human life, because the ecological system can be badly damaged beyond the point where human life can actually be supported effectively. Buddhist economics takes ecological balance into consideration (Puntasen, 2014). Buddhist economics realized the interdependent of human with other beings and nature, so each person's well-being depends on others' well-being and the well-being of earth (Brown & Zsolnaf, 2018). Wise consumption, as noted in Payutto (1992, p. 41), refers to moderation of consumption and overall balance of economic activities with benefits all members of society; that is, consumer realizes that they are a member of society and a part of nature.

Thus, this moderation in consumption (with happiness) is different from conventional economics' utility concepts that seem to imply that the higher consumption individuals have, the higher happiness and satisfaction they gain (by not taken the well-beings of others and nature into account). That conventional thinking model has led to higher consumption and production leading to extensive use of natural resources and pollution creation in a way that conflicts with sustainable development. Aiming to stimulate growth, conventional economics has not questioned or tried to control un-necessary wants/consumption/production, leading to resource exploitation and unsustainable development. In contrast, with BSD, people can be happy in

---

[8] Real quality of life based on Buddhist concept refers to *real need* for life (such as basic needs: food, clothing, shelter, and medicine). Thus, the concept may be different from Western concept of *quality of life*.

the middle-way of life and middle-way economics,[9] implying the realization of true well-being and the balance with other beings and nature.

In sum, the BSD or Buddhist-based concept of sustainable development is different from conventionally used concepts of sustainable development in that it focuses on human behavior that creates problems, while existing paradigms are not concerned with human desires or the root cause of the problem. Most existing approaches and tools (i.e., those in environmental economics) concern market mechanisms, government regulations, and technological advancement in managing resources which are on the supply-side. Thus, the missing or gap so far is the neglect in demand-side management. Therefore, it is essential that this sustainable development path is driven by inner happiness of individuals that eliminates unnecessary wants or defilements place less demand on our environment. When people are happy with the feeling of contentment with what one has and reduce unlimited wants, a lot of natural resources can be saved. Thus, an inner-human focus of happiness development is essential for effective sustainable development.

Kittiprapas (2020) shows the strong relationship of inner happiness indicators such as feelings of having *enough* and happiness in a community case study in Thailand. Contentment with materials significantly affects the happiness of community members. In another case study, (Kittiprapas, 2018b) found that giving/helping/donation is a significant variable to happiness, especially among those who have more ability to give (not in economic suffering). In addition, lower material wants (beyond basic needs) can lead to more happiness for unclassified groups. These case studies support the BSD concept argument that increased happiness does not necessary to go along with the increase material accumulation and wants.

To conclude, BSD is derived from the Buddhist concept of human mind development, assuming that human beings have the capacity to strive for greater happiness with less exploitation, leading to a sustainable path of development as well as living in harmony and peace. Kittiprapas (2016) suggests to define BSD as:

> Buddhist Sustainable Development driven by human-mind development with inner happiness, resulting in moderate consumption and less exploitation of natural resources as well as increased compassion towards humans, other beings and nature (p. 108).

4.6 Looking Forward

As introduced, BSD offers an alternative to existing sustainable development framework by focusing on inner happiness as a key driving force to achieving sustainable development goals. This may be an important link between happiness and sustainable development issues, which have not been addressed during the slow progress of global sustainable development goal. This chapter provides a Buddhist perspective in pursuing sustainable development with a different approach by limiting the demand-side and have different development implications from existing economics

[9] This term is mentioned in Payutto (1992).

and sustainable development frameworks. The chapter discusses related approaches considering happiness and sustainable development aiming to point out existing gaps or address some joint concerns on what is missing in contemporary frameworks. These discussions also remind us to reconsider some traditional values that have been neglected and perhaps that is why the world is in threat of being unsustainable. While sustainable development has not been effectively achieved and has become a global agenda, it may be a time to rethink frameworks. We can explore what has been neglected, the root causes, and how to find a way forward. Though there are some detailed differences, many works discussed in this paper are complementary and increasingly concerned with turning back to traditional values of a good life and inward-looking approaches for change that can bring about real happiness and sustainability. Perhaps, this is a possible direction all perspectives share in common to move forward.

References

Berejnoi, E., Messer, D., & Cloutier, S. (2020). Cultivating spiritual well-being for sustainability: A pilot study. *Sustainability (switzerland), 12*(24), 1–17. https://doi.org/10.3390/su122410342

Brown, C., & Zsolnai, L. (2018). Buddhist economics: An overview. *Society and Economy, 40*(4), 497–513. https://doi.org/10.1556/204.2018.40.4.2

Cloutier, S., & Pfeiffer, D. (2015). Sustainability through happiness: A framework for sustainable development. *Sustainable Development, 23*(5), 317–327. https://doi.org/10.1002/sd.1593

Cloutier, S., Ehlenz, M. M., & Afinowich, R. (2019). Cultivating community wellbeing: Guiding principles for research and practice. *International Journal of Community Well-Being, 2*(8), 277–299. https://doi.org/10.1007/s42413-019-00033-x

Kajikawa, Y. (2008). Research core and framework of sustainability science. *Sustainability Science, 3*(2), 215–239. https://doi.org/10.1007/s11625-008-0053-1

Kittiprapas, S. (2015a). *Achieving sustainable development by inner happiness: A new approach of sustainable development [Presentation]*. Rangsit University.

Kittiprapas, S. (2015b). Buddhist approach and happiness for sustainable development. *The Journal of International Buddhist Studies College, 1*(1), 90–123.

Kittiprapas, S. (2015c). Inner happiness: The key for Buddhist sustainable development. In T. Puntarigvivat & S. Duangsamosorn (Eds.), *Buddhist studies in contemporary world* (pp. 167–203). World Buddhist University.

Kittiprapas, S. (2020). Happiness determinants in a Buddhist community: Where inner happiness matters. *Thammasat Review of Economic and Social Policy, 6*(1), 84–134.

Kittiprapas, S. (2016). *Buddhist sustainable development through inner happiness*. International Research Associates for Happy Societies (IRAH). http://www.happysociety.org/downloads/index/9502/HsoDownload.

Kittiprapas, S. (2017). *Buddhist economics: A holistic development approach for true well-being*. International Research Associates for Happy Societies (IRAH). https://www.happysociety.org/uploads/HsoDownload/9503/download_file.pdf.

Kittiprapas, S. (2018a). *Buddhist sustainable development: A new approach in happiness and development* [e-book]. Thammasat University Press.

Kittiprapas, S. (2018b). What really determine happiness of different groups: A case study of Bangkok. *Journal of Business and Economics, 9*(5), 318–329.

Kjell, O. N. E. (2011). Sustainable well-being: A potential synergy between sustainability and well-being research. *Review of General Psychology, 15*(3), 255–266. https://doi.org/10.1037/a0024603

Krekel & MacKerron. (2020). How environmental quality affects our happiness. In J. Helliwell, R. Layard, J. Sachs, & J. -E. De Neve (Eds.), *World happiness report 2020.*

Maslow's hierarchy of human needs https://5touche.wordpress.com/2017/09/17/5-stages-of-maslows-hierarchy-of-needs/.

Mill, J. S. (1863). *Utilitarianism, liberty, representative government*, 7–9.

O'Brien, C. (2008). Sustainable happiness: How happiness studies can contribute to a more sustainable future. *Canadian Psychology/psychologie Canadienne, 49*(4), 289–295. https://doi.org/10.1037/a0013235

Payutto, P. A. (Phra Debvedi). (1992). *Buddhist economics: A middle way for the market place.* (Dhammavijaya & B. Evans, Trans., 2nd Ed.). Buddhadamma Foundation. http://enlight.lib.ntu.edu.tw/FULLTEXT/JR-MISC/misc140498.pdf.

Payutto, P. A. (1994, December 20). *Peace through Freedom and Happiness (Speech).* 1994 UNESCO Prize for Peace Education.

Payutto, P. A. (Phra Brahmagunabhorn). (1995). *Perfect life* (in Thai, 2nd ed.). Pimsuay.

Payutto, P. A. (Phra Dhammapidok).(2006). *Sustainable Development (in Thai).* Komon Keemthong Foundation.

Payutto, P. A. (Phra Brahmagunabhorn). (2011). *A brief introduction to the Buddha-dhamma* (B. Nirodho & M. Seeger, Eds. &Trans.). http://www.buddhistelibrary.org/en/albums/userpics/10005/a_brief_introduction_to_the_buddha_dhamma.pdf.

Payutto, P. A. (Phra Brahmagunabhorn). (2012). *Buddha Dhamma -Extended Edition* (in Thai). Mahachulalongkornrajavidyalaya University.

Payutto, P. A. (2013). *Peace through Freedom and Happiness.* Wat Nyanavesakavan (forth printed).

Puntasen, A. (2001). *Buddhist economics: Evolution, theory and application in economics* (in Thai). Ammarin.

Puntasen, A. (2014). Buddhist economics and ecology: A Lesson for the future of the asean community. In R. Meinhold (ed.), *Ecological, social and economic sustainability: Perspective for Thailand and ASEAN*. published by Konrad-Adenauer-Stiftung.

Puntasen, A. (2018). From wealth to wellbeing and finally Nibbtina: A transcendence from traditional to buddhist economics. *The Chulalongkorn Journal of Buddhist Studies, 12*(1–61).

Sachs, J. D. (2013). Restoring virtue ethics in the quest for happiness. *The World Happiness Report 2013* (80–95).

Sangasumana, P. (2019). The Buddhist perspective on measuring wellbeing and happiness in sustainable development. *International Journal of Multidisciplinary Research and Development, 6*(3), 243–248.

Sustainable Development Solution Network. (2013, October). An action agenda for sustainable development: Report for the UN Secretary-General.

The World Commission on Environment and Development (The Brundtland Commission). (1987). *Our common future.* Oxford University Press.

Ulluwishewa, R. (2016). Spirituality, sustainability and happiness: A quantum-neuroscientific perspective. In S. Dhiman & J. Marques (Eds.), *Spirituality and sustainability: New horizons and exemplary approach* (pp. 155–168). Springer.

Wamsler, C., Brossmann, J., Hendersson, H., Kristjansdottir, R., McDonald, C., & Scarampi, P. (2018). Mindfulness in sustainability science, practice, and teaching. *Sustainability Science, 13*(1), 143–162. https://doi.org/10.1007/s11625-017-0428-2

Zidansek, A. (2007). Sustainable development and happiness in nations. *Energy, 32*(6), 891–897.

Chapter 5
Mahatma Gandhi's *Sarvodaya* (Welfare for All) as a Way to End Violence and Achieve Happy, Sustainable Societies

Jorge Guardiola Wanden-Berghe

Abstract I argue that suffering and violence is prevalent in the world, and that most people have adapted to it, reflecting what Arendt (Eichmann in Jerusalem: A report on the banality of evil. Viking Press, 1964) referred to as *the banality of evil*. To prevent suffering, we must develop the idea of *good* happiness. I understand good happiness to be directly linked to *Sarvodaya*, as it entails a person acting to enhance his/her own happiness and simultaneously that of others. I propose a set of guidelines regarding what is needed to embark on the path towards *Sarvodaya*: (1) A better understanding by human beings about their role in the environment, that is, to see themselves as belonging to the land rather than the land belonging to them; (2) Scientific learning from other cultural paradigms that view human beings in this way, such as *Buen Vivir*, Buddhism or *Ubuntu*; (3) Changes to State investments, such as decreasing military expenditure and increasing social, health, education and environmental protection expenditure; (4) Satisfaction of human needs rather than material desires, which requires a paradigm shift away from planned and perceived obsolescence.

Keywords Sarvodaya · Happiness · Sustainability · Violence · Ethics

5.1 Introduction

In her famous book *Eichmann in Jerusalem: A report on the banality of evil*, Hanna Arendt (1964) argued that Adolf Eichmann, one of the architects of the Holocaust, did not commit his crimes against humanity because he was a psychopath. Instead, she argued that Eichmann was a normal person, and that the holocaust was generally carried out by normal people acting in a way she referred to as *the banality of evil*. That is, people were capable of inflicting extreme violence on others in the belief that it was the normal thing to do under certain circumstances. Indeed, many people

J. G. Wanden-Berghe (✉)
Institute of Peace and Conflicts, Department of Applied Economics, Universidad de Granada, Granada, Spain
e-mail: jguardiola@ugr.es

S. Cloutier et al. (eds.), *Linking Sustainability and Happiness*, Community Quality-of-Life and Well-Being,
https://doi.org/10.1007/978-3-030-89559-4_5

committing violence may be living an ordinary, happy life, disengaged from the consequences of their life choices.

Recent statistics on happiness show that, in many countries, the majority of the population consider themselves to be happy. In fact, the proportion of citizens dissatisfied with their lives exceeds 40% in only 11 countries (World Values Survey, 2019), a situation that coincides with economic, social and environmental problems. These high average happiness levels coexist with social and environmental problems that bring suffering to sentient beings. These are problems we are facing now and, depending on the chosen solution, will exist in the future to a certain degree. It seems that, after centuries of periods of peace and violence, we have learned to lead happy lives while coexisting with different types of violence in the world, creating a norm where violence is commonplace or, in Arendt words, a banality of evil.

Violence is present in many forms that are normalized in society, or even fostered by some groups. An important example is the refugee and immigrant crisis the world is facing now, with people deciding to vote for leaders who seek to build physical and/or legal walls to prevent immigrants from safely reaching and living in the US and Europe. There are extreme cases, such as the situation in Italy in 2018, when authorities denied NGOs permission to help people who were dying in the ocean in an attempt to emigrate from Africa to Europe, and even prosecuted those people who tried to save migrants' lives. The fact that people enjoy different rights just because of where they were born or the color of their skin is also a type of violence that infringes on human rights.

Another case of normalized violence, far less visible than the previous one, is the production culture in capitalism, particularly planned obsolescence and perceived obsolescence. Perceived obsolescence increases consumption through advertisements and publicity seeking to foster and create desires, and trap people on a hedonic treadmill; that is, people are driven to constantly buy products and services that only briefly increase their happiness (Kasser, 2002). Planned obsolescence is a supply-side philosophy of production: the aim is to produce goods that do not last long so that they will soon have to be bought again. While planned obsolescence and perceived obsolescence increase companies' profits, they have high costs as they condemn humanity to a life of working and consuming that does not bring happiness.[1] They are also responsible for modern slavery, as manufactured products such as clothes and electronic devices, and the extraction of their components, often require cheap, intense labor by people in the Global South, with no health protection or labor rights (a model that is a heritage of slavery and colonialism). In general, both types of obsolescence types destroy nature and condemn people to lead unhappier lives because of the contamination of the environment. There is a direct negative relationship between happiness and pollution (Apergis, 2018), as well as an indirect impact through the

[1] In fact, empirical evidence shows that time spent in relationships is a good predictor of happiness (Becchetti et al., 2011), and a culture that encourages people to dedicate time to consumption and working diminishes the possibility of spending quality time with friends and relatives.

effects of natural disasters such as floods (Luechinger & Raschky, 2009). Furthermore, these problems will be worse for the children and grandchildren of the present generation.

The food industry that enslaves animals is another example of violence that is widely tolerated. In general, humans treat animals as commodities, causing them severe harm in order to satisfy the increasing demand for meat and generate ever-greater private profits. On the one hand, eating meat leads to a (very debatable) improvement in people's nutrition as a possible benefit, but on the other hand it damages the environment and the land, and causes enormous animal suffering.

In consequence, over many centuries, we have learned to normalize violence in contexts of peace in a globalized world where the internet allows easy access to information. We have even managed to achieve a happy life in an imperfect equilibrium with rising inequalities and tensions. The world left to our children and grandchildren will be one with ever more frequent human-induced climate disasters and potential conflicts over resources (Salehyan, 2008). The question of how this problem can be solved is difficult to answer.

A possible way for academia to tackle the problem is to visualize the sources of violence that coexist with peace in the world, those that remain invisible or unacknowledged, and propose solutions to overcome them. This is the path taken in this chapter, where I discuss how to foster happiness and environmental sustainability by peaceful means. To that end, I adopt a modern version of the Gandhian definition of *Sarvodaya* (López Martínez, 2017), which means welfare for all; that is, not only for the greatest number possible (as in the utilitarian perspective of Jeremy Bentham) but for everyone. With this concept, Gandhi (1977) proposed establishing institutions in society that guarantee everyone's wellbeing, as well as fostering values such as equality, justice and solidarity through direct participation. I take the *Sarvodaya* idea further to include future generations (humans and other living beings): they are not present now but they deserve a good quality of life in the future, even though they cannot democratically participate in the present. Therefore, in this broad conception of *Sarvodaya*, the environment plays a prominent role, as our love of life and the wellbeing of future generations who are going to live on our planet requires us to respect the environment. There are modern concepts that aim to capture the happiness of current generations, which are closely aligned with the *Sarvodaya* concept: for instance, *sustainable happiness* is defined by O'Brien (2008) as "the pursuit of happiness that does not exploit other people, the environment, or future generations" (p. 289). However, *Sarvodaya* refers more to directly enhancing happiness in a collectivist way, putting the emphasis on meeting everyone's needs and not just avoiding exploitation. Therefore, I argue that a necessary paradigm for a happy and sustainable world should include the satisfaction of all people's needs. Gandhi (1977) believed that *Sarvodaya* could only be achieved through direct and nonviolent participation in public decisions, and democratic control of resources that are necessary for people to satisfy their human needs (López Martínez, 2017). *Sarvodaya* also involves giving power to the powerless; that is, people who are poor and marginalized, and who face the greatest obstacles to achieving wellbeing.[2]

In this chapter, the *Sarvodaya* concept serves as the basis for a discussion on an ethical understanding of happiness oriented towards the satisfaction of the needs of all humans (present and future generations). Specifically, my aim is to introduce a pacifist understanding of happiness from a Gandhian perspective, drawing inspiration from peace research and ecological economics. I build on some of the literature on happiness, ecological economics, and peace studies, and discuss some ideas that may enrich our understanding of the link with those topics. The chapter is the product of my own understanding of peace, happiness, and nature; as such, it is a discussion chapter based on an ideological concept grounded in pacifism and does not yield any replicable results. All disciplines can be enriched by learning from other disciplines, so the disciplines from which this chapter draws its inspiration may also be enriched by evidence from happiness research. The common denominator in this chapter is the *Sarvodaya* concept, which enables a better ethical understanding of happiness; that is, it allows us to differentiate between "bad" happiness, "neutral" happiness, and "good" happiness, in relation to the impact that pursuing happiness may have on other people's happiness. I propose that *Sarvodaya* should be related with "good" happiness. The "bad" type of happiness should be conceptually related to the happiness experienced by an individual who, in their pursuit of happiness, causes unhappiness in others, reflecting the idea of the banality of evil.

To achieve this objective, we divide this chapter in standard scientific sections (intro, methods, results, discussion and conclusions). In the methods section I include a brief paragraph on how I found my sources. In the results I present the notion of violence in peace theory, using some illustrative examples, and I show some empirical evidence on the normalization of violence. In addition, I propose several paths to follow to bring culture in line with *Sarvodaya*, instead of responding to the greedy interests of the very few, which in turn creates violence. In this vein, I propose an ethical understanding of happiness oriented towards *Sarvodaya* in order to foster other people's happiness and encourage respect for the environment. To that end, I argue that what is needed is a better understanding of human beings—learning from other cultural paradigms that see the components of reality as interconnected—good public policies, and a focus on satisfying human needs instead of material desires. The last sections are devoted to discussing the limitations, research implications and policy proposals, and drawing some conclusions.

5.2 Methods

This is a discussion chapter, therefore there is no possibility of replication of the results obtained. In fact, it is based on the opinion of the author, and given that this opinion is subjective, it serves to reflect from a philosophical basis on the topics

[2] In the India of Gandhi's time, and in the present, these people were specially the untouchables, the lowest caste in India's caste system. Gandhi called them *harijans* (sons of God).

of happiness, violence and sustainable development. The sources are mainly bibliographical references, and those were found during several years of study on the topics mentioned. Additional sources were found looking in repositories such as Google Scholar.

5.3 Results

5.3.1 What is Violence?

5.3.1.1 A Taxonomy of Violence

The activism of the Nobel peace prize winner Martin Luther King Jr. was based on three pillars: the fight against racism and oppression, the fight against poverty, and opposition to the military intervention in Vietnam. All three are examples of the types of violence that comprise the theory presented in this section. According to Johan Galtung (2004), there are three types of violence (Fig. 5.1): cultural, structural and direct violence. The first two types are subtle and often invisible, in contrast with direct violence, which is more evident.

As part of Martin Luther King Jr.'s vision, he aimed to defeat racism as a source of cultural violence (Schraff, 2008). Cultural violence refers to a particular way in which we think and see the world, which involves violence. In addition to racism, other sources of cultural violence are male chauvinism, extreme selfishness, or aporophobia. This type of violence can be understood as the opposite of *Sarvodaya,* as cultural violence runs counter to the values that foster welfare for all.

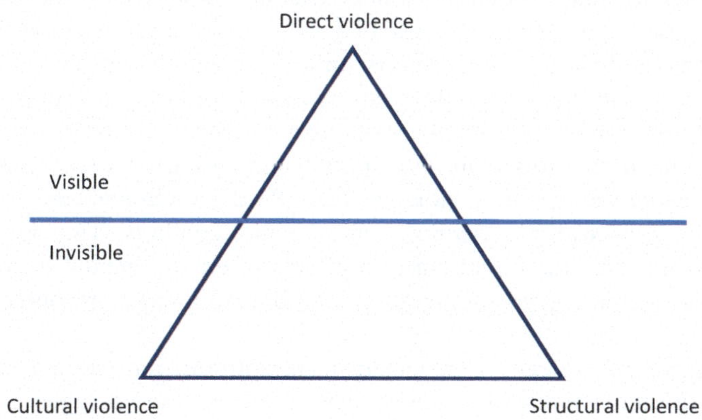

Fig. 5.1 Three types of violence

Cultural violence is a major cause of the other two types of violence. Poverty and income inequality, and the norms and social institutions that shape them, are important components of structural violence. Poverty and inequality are closely associated with cultural values that tolerate it (the marginalization of certain people and opposition to the distribution of wealth, recognition of human rights, or coexistence with other cultures). Official statistics point to a reduction in inequality since 1988, while other research that takes into account geopolitical differences and the effect of China suggests that inequality has tripled since 1960 (Hickel, 2017).

Values such as greed and selfishness create a production structure that steers production and consumption towards what people desire, instead of what they need. Mahatma Gandhi captured this imbalance in a famous quote: "earth provides enough to satisfy every man's need but not every man's greed" (1977, p. 32). By creating infinite and insatiable desires, planned and perceived obsolescence play a major role in cultural violence, which ultimately generates structural violence. For instance, they contribute to environmental degradation, which is another component of structural violence. Environmental degradation is related to inequality, as the harms of climate change are, and will continue to be, unevenly distributed: those that pollute more suffer less harm while those polluting less suffer the most. According to an Oxfam report (Gore, 2015), the poorest half of the world's population—approximately 3.5 billion people—generates only about 10% of total global emissions attributed to individual consumption, yet they mostly live in the countries that are most vulnerable to climate change. In contrast, about 50% of these emissions can be attributed to the richest 10% of the world's population, whose average carbon footprint is up to 11 times that of the poorest half of the population, and 60 times that of the poorest 10%.

Direct violence consists of insulting, physically harming or killing others, taking its most extreme form in war. A major cause of direct violence is cultural and structural violence. Martin Luther King Jr. believed that people's tolerance of war allowed politicians' intervention in Vietnam (Schraff, 2008). This tolerance of war, or seeing the other as an enemy because of a different ideology, is cultural violence. Of course, structural violence by means of military expenditure enables wars, which become bloodier as military technology advances and military expenditure increases. Harming the environment is an example of direct violence towards living beings and their future generations, and ideas about production (cultural violence) and overexploitation of resources (structural violence) feeds the destruction of land, ecosystems, and biodiversity, as well as creating environmental conflicts. Therefore, cultural violence also includes the direct harmful effects of environmental degradation in terms of poor health, forced immigration, conflict over scarce resources and lost lives.

The examples of violence I presented in the introduction can also be categorized under the three types of violence. The immigrant and refugee crisis is fed by racism and aporophobia (cultural violence), which creates marginalization and poverty as immigrants and refugees have fewer rights than citizens in many countries, and racism prevents them from finding a decent job (structural violence). Racism, in turn, can lead to aggressions against these groups of people (direct violence).

The cultural violence of consumption, planned obsolescence and perceived obsolescence leads to the exploitation of other people and the environment as a structural phenomenon (structural violence), preventing those people from enjoying their free time, leading healthy lives, getting an education, enjoying nature, or living from what nature provides, as well as making them more vulnerable to natural disasters (direct violence). Thinking that animals exist for human consumption or denying that they have feelings is cultural violence, which causes the structural violence embedded in the food industry, involving animal exploitation and suffering. These activities also damage the environment; for instance, through emissions from cows and soil degradation from cattle grazing (direct violence).

5.3.1.2 The Banality of Evil

Evidence of the normalization of violence is the high levels of happiness coexisting with violence. According to the 2010–2014 wave of the World Values Survey (2019), the percentage of people worldwide who are dissatisfied with life—that is, people who choose 5 or less on a 1–10 scale to value their life satisfaction—is 27%, or a total of 89,565 individuals from 60 countries. National levels of dissatisfaction with life range from 7% in the Netherlands to 59.1% in Armenia. Dissatisfaction exceeds 40% in only 11 countries, which belong to the South. In contrast, in 33 countries the proportion of dissatisfied people is lower than 25%. This seems to indicate high levels of happiness generally, despite the obvious differences between countries. Further empirical research is needed on this point, but a possible explanation relates to the banality of evil in society; that is, people manage to be happy while coexisting with violence.

Regarding the normalization of violence, Arendt's hypothesis of the banality of evil could be seen as out of date. Indeed, her book dates from 1964 and makes reference to events that took place in a context of war. It could thus be argued that the context may have triggered the inhuman treatment witnessed in the holocaust, which is difficult to extrapolate to other scenarios. However, experimental tests undertaken outside a war context, some of them quite recent, point to the existence of a tendency to engage in potentially violent acts if someone is in a hierarchy, or if the context requires it. The well-known Milgram (1965) experiments involved a volunteer, acting as the teacher, administering electric shocks to an actor, playing the role of learner. Unknown to the "teacher," the electric shocks were fake, but he had to administer them to the "learner" every time the latter answered one of the former's questions incorrectly. The volunteer administered the electric shocks because he was required to do so according to the instructions of the researcher conducting the experiment, who urged the teacher to administer the shocks each time he hesitated. The teacher could not see the learner but he could hear the faked screams of pain each time he administered a shock. The experiment included only men, but a recent replication included women (playing the part of the teacher and the learner) and found similar results (Doliński et al., 2017). In addition, in the infamous Stanford prison experiment in 1971 some volunteers playing the role of guards in a mock prison ended up abusing the volunteers playing the role of prisoners. These experiences reflect Hannah Arendt's vision of the banality of evil, so in this chapter I use the concept of the banality of evil as synonymous with the normalization of violence.

Given that all violence brings suffering, and the roots of all of violence lie in cultural violence, I propose in the next section a change in our understanding needed to embark on the path to *Sarvodaya*.

5.3.2 Transforming Cultural Violence into Sarvodaya

In the previous section, I argued that culture and the way we interpret the world is at the heart of violence. In order to create a world in accordance with the *Sarvodaya* idea and diminish cultural violence, it would seem necessary to change our perceptions of the world. This implies becoming conscious of what is going wrong, which in turn entails an awareness of the banality of evil: how have we managed to tolerate cultural, structural, and even direct violence around us and in spite of it still lead happy lives? In my view, this could bring about greater unhappiness in the short term by acknowledging all the suffering in the world, but in the long term it would allow us to advance towards *Sarvodaya*. Therefore, tolerance and empathy for the other living beings in the world, and those that are still to come, could help create fair societies. This means fostering universal values more closely aligned with empathy and loving kindness, instead of selfish ones such as the pursuit of power, the importance of achievement, and hedonism (Schwartz, 1992).

In happiness studies, making progress towards *Sarvodaya* requires an ethical perspective on happiness. An ethical understanding of happiness that benefits all sentient beings should be identified and enhanced. There is a need for deep reflection on this issue, but I suggest that a starting point could be to categorize people's actions in pursuit of happiness and the prevention of suffering as "good," "bad" or "neutral," according to the effects of those actions on others. Bear in mind that I refer to effects on other people and animals that exist now and also those that will exist in the future; therefore, environmental preservation plays an important role. I refer to action in a broad sense, including motivations and livelihoods, and also thoughts, words and feelings experienced by a person. In this framework, "good" happiness involves actions done by a person that help increase that person's happiness while also increasing other people's happiness, in line with the concept of *Sarvodaya*. On the contrary, "bad" happiness would refer to actions that increase the happiness of the person doing them, while harming other people and/or the environment, thereby creating violence and fostering its normalization. In other words, "bad" happiness arises when people enhance their own happiness at the expense of others, who suffer from the different forms of violence outlined in the previous section. Lastly, "neutral" happiness would refer to actions such as moving our head, which may slightly affect happiness in the person performing that action but without affecting others. In my understanding, everything is interconnected, therefore very few "neutral" actions exist. In order to overcome cultural violence and create sustainable and happy societies, the emphasis should be put on enhancing "good" happiness to achieve *Sarvodaya.*

Certainly, there is a need for more in-depth research on those issues, to answer questions such as: What actions respect (or at least cause minimal harm to) the environment? What makes people happy while making others happy or unhappy? This could help build alternatives to "bad" happiness, which involves the three types of violence. In the first line of research (how to be happy while respecting the environment), some attempts have been made to identify how our actions impact the environment while aiming to satisfy people's needs (Creutzig et al., 2018; Guillén-Royo, 2016; O'Neill et al., 2018; Steinberger & Roberts, 2010), but much research is still needed. In the second line, there are some studies that suggest that, with a few specific exceptions, actions benefiting others (Dunn et al., 2008, 2014) and benefiting the environment (Binder & Blankenberg, 2016; Kasser, 2017; Suárez-Varela et al., 2016) increase happiness. This scientific evidence can be a good starting point for discussion.

The key question is how to make the shift from happiness coexisting with cultural violence to a "good" happiness. Below, I propose some guidelines on how we can advance conceptually in this way.

5.3.2.1 A Better Understanding of Human Beings' Role in the Environment

In the line with the *Sarvodaya* concept, it is necessary to adopt a vision of humanity in a state of interconnectedness with nature, other living beings and other humans, in order to help achieve welfare for all. This is a vision committed to understanding the world in terms of sustainability, and is directly opposed to other harmful paradigms that view people as individuals detached from this reality, such as the concept of homo oeconomicus (Urbina & Ruiz-Villaverde, 2019). Indeed, this vision would help to construct an idea of "good" happiness.

There have been recent attempts to examine different conceptions of the self as part of the natural world. For instance, Hirvilammi and Helne (2014) and Helne and Hirvilammi (2015) use the concept of ecological self developed by Arne Naess in the 1970s. According to this concept, each person is embedded in a complex web of relationships with other human beings and with nature. Therefore, nature is not at the service of people, but rather people are part of nature (Naess, 2009). Arne Naess was a pioneer of deep ecology and he understood human beings from a holistic perspective as a part of the natural world. Other philosophers before him also expressed similar views. For example, Alfred Whitehead, in his process-relational philosophy, also saw the world as a web of interrelated processes, considering humans as an integral part, responsible for their actions, which have consequences in the world. The process-relational philosophy also influenced John Cobb, who co-authored with Herman Daly an important work on ecological economics (Daly & Cobb, 1994).

As we can see, there are great thinkers from the North with great ideas for gaining a better understanding of people in order to achieve welfare for all. A very valuable view of happiness comes from experiences in the South, such as the *Buen Vivir* (Good Living) indigenous philosophy in the Andean region, *Ubuntu* in South Africa and Buddhist practice in Asia. Those ideas, particularly Buddhism, started to spread in the North during the second half of the twentieth century. What these perspectives have in common is an understanding of the self as being interdependently bound to other life forms.

Empirical evidence on happiness shows that there is a positive correlation between a holistic understanding of the environment and being happy. For instance, it has been found that indigenous Andean people in a rural area in Ecuador who follow the *Buen Vivir* philosophy have high levels of happiness (García-Quero & Guardiola, 2018; Guardiola & García-Quero, 2014). In addition, a questionnaire about people's relatedness to nature given to a sample of students and business executives in Canada showed that people experienced greater happiness when they were more connected to nature (Nisbet et al., 2011). This questionnaire included statements such as "My connection to nature and the environment is a part of my spirituality" to which participants had to indicate their agreement using a Likert scale.

5.3.2.2 Learning from Other Cultural Paradigms

Related to the previous point, it must be understood that other ways of development are possible and that some of those are, in fact, more ecologically friendly and more conducive to happiness than those embedded in capitalism. These alternative cultural paradigms are closer to *Sarvodaya* than the cultural violence paradigms that form part of other organizational systems, and in the societies where they occur we see a normalization of the counterparts of violence.

This is the case with the indigenous philosophy of *Buen Vivir*, which was popularized in Ecuador and Bolivia when progressive governments seeking to visibilize indigenous cultures put this way of life at the center of their policies (Hidalgo-Capitán & Cubillo-Guevara, 2014). Before then, however—indeed, ever since colonization—indigenous people had been marginalized, even though their perspective of life was closer to *Sarvodaya* than the perspective of their colonizers and subsequent oppressors. A similar argument can be made for indigenous groups in most parts of the world (de Sousa Santos, 2010; Said, 1979). In sum, many people that have been discounted as poor, savages or underdeveloped hold the cultural key for transformation, as their cultural understanding of life is inclusive of the environment and other human beings.

We can learn from the *Buen Vivir* paradigm thanks to the researchers that have studied it, from both the South (Acosta, 2012; Gudynas, 2011) and the North (García-Quero & Guardiola, 2018; Tortosa, 2009). Buddhism has also been popularized, and an ecological economics perspective has been incorporated, placing Buddhism as an alternative way of life or alternative to development theories through the so-called Buddhist economics (Schumacher, 1973). Buddhism has traditionally built on a dualistic interpretation of the world that leads to ignorance, as it holds that everything is interconnected. That is, any phenomenon in the world has the characteristic of emptiness, and any single thing in the world exists in relation to everything else. This provides a powerful insight into holism that can be very useful when it comes to encouraging a paradigm shift towards *Sarvodaya*. These ideas still need to reach the general public and be translated into new actions (in the broad sense defined above).

5.3.2.3 Good Investment by the State

If institutions in society aim to achieve welfare for all, especially for the most disadvantaged, then a wise use of the public budget should be directed towards it. Recent research has suggested that happiness could be a public policy goal (Stiglitz et al., 2009), and has identified factors that increase global happiness (Di Tella et al., 2003). It has been empirically proven that hunger (Rojas & Guardiola, 2017), environmental degradation (Di Tella & MacCulloch, 2008; Welsch, 2002) and failure to meet human needs in general (Guillén-Royo et al., 2013) lead to unhappiness, so it seems clear that public authorities should seek to end hunger, respect the environment and permit people to satisfy their human needs, thus helping to achieve *Sarvodaya*.

It is therefore necessary to direct government budget towards achieving welfare for all. However, it should be borne in mind that achieving welfare for all is quite different to the maximization of welfare proposed by some economic models. It seems that government social expenditure increases citizens' happiness (Di Tella et al., 2003; Perovic & Golem, 2010). In fact, the money devoted to increasing happiness for the most disadvantaged would not only be desirable in terms of *Sarvodaya*, but may also increase average happiness, as was empirically demonstrated for the case of unemployment subsidies (Di Tella et al., 2003).

Military expenditure is a component of structural violence, whose link with happiness has not been well researched, although a negative relationship would be expected. Spending on weapons and defense has a double harmful effect on quality of life: (1) it enables direct violence through the use of weapons; and (2) it entails an opportunity cost as the money devoted to weapons spending could be redirected to more important budget items such as health and education. Therefore, reducing military expenditure and devoting this money to spending on health and education would yield what is known as the peace dividend, which is the additional welfare gain produced by decreasing military expenditure (Meulewaeter, 2016).

5.3.2.4 Satisfy Human Needs, Not Material Desires

In order to achieve *Sarvodaya* all people's needs should be met, not only material needs but others such as participation, identity, and freedom, among others (Guillén-Royo, 2016; Guillen-Royo et al., 2017). However, society is centered on the satisfaction of the desires of a few, through institutions such as markets, in which only people with money can successfully meet their needs and desires. This old paradigm is linked with the cultural violence of planned and perceived obsolescence, among other cultural norms. This cultural violence translates into structural violence, such

	Means	Ends
Paradigm to overcome	Human resources Natural resources	Companies: Maximize profits People: Pursue money or status
New paradigm	Economic sustainability	*Sarvodaya* (human needs and respect for nature)

Fig. 5.2 Changing paradigms

as poverty and inequality, as well as direct violence such as environment disasters, or wars to control resources like oil. In my view, to achieve welfare for all, production without planned and perceived obsolescence is needed.

The art critic John Ruskin, author of the book *Unto This Last* (1985), had a huge influence on Gandhi's concept of *Sarvodaya.* If fact, Gandhi translated several parts of this book into Gujarati and called the resulting book *Sarvodaya.* Ruskin's focus was the effect of production on people. As he so beautifully argued,

> You may grow for your neighbour, at your liking, grapes or grapeshot; he will also, catallactically, grow grapes or grapeshot for you, and you will each reap what you have sown. It is, therefore, the manner and issue of consumption which are the real tests of production. Production does not consist in things labouriously made, but in things serviceably consumable; and the question for the nation is not how much labour it employs, but how much life it produces. For as consumption is the end and aim of production, so life is the end and aim of consumption (1985, p. 125).

In line with Ruskin's book, I propose that life (a happy life, a good life or a life worth living) should be the end and aim of production and consumption. Therefore, humanity and nature should not be a means but an end in itself, which requires a paradigm shift. Figure 5.2 shows the paradigm to overcome, wherein humanity and nature are seen as resources, while materialistic components of life, such as money or status, are seen as ends. Ends and means have to be switched in the new paradigm; that is, people and nature should be seen as an end in themselves, while money should be seen as a means to achieve it.

5.4 Discussion, Limitations, Further Research and Policy Implications

This chapter is an attempt to shed some light on the intersections among happiness studies, peace studies and ecological economics. It is my aim to present an idea of happiness that is pacifist and inclusive, by using the *Sarvodaya* conceptualization. However, this discussion chapter is not free from limitations, some of which point to future research avenues: Firstly, attempts to revise the term *Sarvodaya* and bring it up to date would benefit from other approaches in future research, especially the feminist perspective. Gandhi's (1977) thoughts and actions and his efforts to give power to the powerless did not center on women's rights. As time has passed, women's empowerment has been bolstered by women's liberation movements. In particular, approaches such as ecofeminism offer a greater understanding of a women's vision to transcend violence into peace (Mies & Shiva, 1993). This philosophy goes beyond the difference between North and South outlined above, and aims to empower women to achieve greater connections and tolerance among nature and human beings, particularly oppressed people.

Secondly, as mentioned above, an ethical perspective of happiness should be rooted in deeper reflection. While *Sarvodaya* is surely not the only useful path towards a conception of "good" happiness, I propose it in this chapter because in my view it is the most suitable. However, inspiration could also be drawn from many philosophers, schools of thoughts and cultures, which may not be so different from *Sarvodaya.*

Thirdly, as this is a discussion chapter, it lacks empirical evidence, which is needed to support the arguments presented here. In particular, the argument about the banality of evil is a plausible hypothesis to test in happiness studies. In this vein, there are several questions that need further empirical evidence, and are left for further research that goes beyond the scope of this chapter: To what extent and under what circumstances does the banality of evil arise in a society? How do different types of violence influence happiness and the environment? How does happiness differ between those that impose violence on others and those that suffer the violence? If people suffering violence claim to be happy anyway, is it because they have adapted to this, because they obtain happiness from other sources, or is there an alternative explanation? Evidence on these matters could help explain the possible banality of evil and pave the road to *Sarvodaya.*

Finally, I built on empirical studies of happiness as a way of understanding how to achieve welfare for all. These studies follow a scientific empirical approach, which is normally more objective than other, more discursive approaches. Nevertheless, we should take the findings of those studies with caution as they do not really reflect welfare for all, but rather report general correlations between wellbeing and other variables. They typically find, *ceteris paribus* (that is, assuming that all the other components of reality remain constant), relationships between happiness and other aspects of life. While the statistical results help to understand reality, they do not show reality in a holistic way. The marginal relationship is considered valid when it is significant, even though the magnitude of the correlation may be very low. In this vein, the explanatory power of the models aimed at examining happiness is normally very low, which means that although correlations are found between happiness and other aspects of life, happiness is not fully understood.

Relating happiness research with an ethical perspective such as *Sarvodaya* and visibilizing violence means political action can be taken. This action can be devoted to raising people's awareness of how much violence has been tolerated in our society, and how it is tolerated, in order to encourage people's direct participation in providing solutions and making institutional changes. In general, moving towards *Sarvodaya* (the "good" happiness idea I propose) requires multiple efforts from multiple parties. In my view, the best hope of achieving welfare for all is through education—an education that includes healthy values, such as empathy and social justice, and enables a transformation of people's lives so they are no longer considered instruments of the market (López Castellano et al., 2018). Critical pedagogy (Freire, 2005) helps people recognize the oppression that the marginalized are suffering, and opens the door to liberating themselves of this oppression and nonviolent empowerment.

Some people may believe that bringing an end to violence and incorporating the *Sarvodaya* perspective are unachievable goals, viewing them as a utopia. In fact, at the beginning of the twenty-first century, many social movements took to the streets in search of this utopia, demanding greater empowerment of disadvantaged people, and real democracy for a better world, in a similar sense to *Sarvodaya.* Examples include movements such as Occupy Wall Street or We Are The 99%, in the US and the UK, 15 M in Spain, or the indigenous marches around Latin America (Tortosa, 2012). Therefore, *Sarvodaya* or any similar project aimed at the betterment of society and the preservation of the environment should be considered not only a goal but a necessary path to follow. The Uruguayan writer Eduardo Galeano reflects this very well in the following lines:

> Utopia lies at the horizon.
> When I draw nearer by two steps,
> it retreats two steps.
> If I proceed ten steps forward, it
> swiftly slips ten steps ahead.
> No matter how far I go, I can never reach it.
> What, then, is the purpose of utopia?
> It is to cause us to advance (1993, p. 230).

5.5 Conclusions

In this chapter, I have discussed how to embark on the path towards a happier world that respects life and the environment. My proposal is influenced by Mahatma Gandhi's concept of *Sarvodaya,* which means welfare for all, not just the few. I have suggested that this concept should encompass not only the living beings that exist now, but also those that will exist in the future, thereby assigning importance to projected environmental problems. It is my understanding that Gandhi did not directly include this dimension as the problems he faced in his times were different from those we face now.

I have pointed to the need to enhance happiness insofar as it contributes to *Sarvodaya*, and to this end I have distinguished between "good," "bad" and "neutral" happiness. The first category of happiness is directly linked to *Sarvodaya*, as it entails a person acting to enhance his/her happiness and simultaneously that of others. On the contrary, "bad" happiness means increasing one's life satisfaction through the suffering of others. The suffering of others is, I argue, normalized into cultural, structural and direct violence, which we may perceive as normal. The challenge is therefore to be guided by *Sarvodaya,* or any idea of "good happiness," in order to avoid what Hannah Arendt termed "the banality of evil"—that is, adapting to and accepting the violence involved in "bad" happiness. Guidelines for achieving *Sarvodaya* are related to a better understanding of life that is open to other cultures and that leads to action aimed at satisfying the needs of all humans, both those existing now and those that will exist in the future.

Acknowledgements I would like to thank Mònica Guillén Royo and Fernando García Quero for their comments and suggestions about this research. A special thanks to José María Tortosa, without whose continuous support and guidance the most daring thoughts in this paper would not have been possible. An anonymous reviewer of this chapter made very useful comments that helped to improve it. Of course, I bear sole responsibility for any errors there may be in this chapter.

Funding The author acknowledges the financial support from the Spanish *Agencia Estatal de Investigación* and the *European Regional Development Fund* (project ECO2017-86822-R); the *Regional Government of Andalusia* and the *European Regional Development Fund* (projects P18-RT-576 and B-SEJ-018-UGR18) and the University of Granada (*Plan Propio. Unidad Científica de Excelencia: Desigualdad, Derechos Humanos y Sostenibilidad -DEHUSO-*).

References

Acosta, A. (2012). *Buen Vivir/SumakKawsay. Una oportunidad para imaginar otros mundos*. Abya Yala.

Apergis, N. (2018). The impact of greenhouse gas emissions on personal well-being: Evidence from a panel of 58 countries and aggregate and regional country samples. *Journal of Happiness Studies, 19*(1), 69–80. https://doi.org/10.1007/s10902-016-9809-y

Arendt, H. (1964). *Eichmann in Jerusalem: A report on the banality of evil*. Viking Press.

Becchetti, L., Trovato, G., & Bedoya, D. A. L. (2011). Income, relational goods and happiness. *Applied Economics, 43*(3), 273–290. https://doi.org/10.1080/00036840802570439

Binder, M., & Blankenberg, A. K. (2016). Environmental concerns, volunteering and subjective well-being: Antecedents and outcomes of environmental activism in Germany. *Ecological Economics, 124*, 1–16. https://doi.org/10.1016/j.ecolecon.2016.01.009

Creutzig, F., Roy, J., Lamb, W. F., Azevedo, I. M. L., Bruine de Bruin, W., Dalkmann, H., Edelenbosch, O. Y., Geels, F. W., Grubler, A., Hepburn, C., Hertwich, E. G., Khosla, R., Mattauch, L., Minx, J. C., Ramakrishnan, A., Rao, N. D., Steinberger, J. K., Tavoni, M., Ürge-Vorsatz, D., & Weber, E. U. (2018). Towards demand-side solutions for mitigating climate change. *Nature Climate Change, 8*(4), 260–263.https://doi.org/10.1038/s41558-018-0121-1

Daly, H. E., & Cobb, J. B. (1994). *For the common good: Redirecting the economy toward community, the environment, and a sustainable future*. Beacon Press.

de Sousa Santos, B. (2010). *Descolonizar el saber, reinventar el poder*. Ediciones Trilce.

Di Tella, R., & MacCulloch, R. (2008). Gross national happiness as an answer to the Easterlin Paradox? *Journal of Development Economics, 86*(1), 22–42. https://doi.org/10.1016/j.jdeveco.2007.06.008

Di Tella, R., MacCulloch, R. J., & Oswald, A. J. (2003). The macroeconomics of happiness. *Review of Economics and Statistics, 85*(4), 809–827. https://doi.org/10.1162/003465303772815745

Doliński, D., Grzyb, T., Folwarczny, M., Grzybała, P., Krzyszycha, K., Martynowska, K., & Trojanowski, J. (2017). Would you deliver an electric shock in 2015? Obedience in the experimental paradigm developed by Stanley Milgram in the 50 years following the original studies. *Social Psychological and Personality Science, 8*(8), 927–933.

Dunn, E. W., Aknin, L. B., & Norton, M. I. (2008). Spending money on others promotes happiness. *Science, 319*(5870), 1687–1688. https://doi.org/10.1126/science.1150952

Dunn, E. W., Aknin, L. B., & Norton, M. I. (2014). Prosocial spending and happiness. *Current Directions in Psychological Science, 23*(1), 41–47. https://doi.org/10.1177/0963721413512503
Freire, P. (2005). *Pedagogy of the oppressed*. Continuum.
Galeano, E. (1993). *Las palabras andantes*. Siglo XXI.
Galtung, J. (2004). Violence, war, and their impact: On visible and invisible effects of violence. *Polylog: Forum for intercultural philosophy 5*. http://them.polylog.org/5/fgj-en.htm
Gandhi, M. K. (1977) Hacia un socialismo no violento. La Pléyade.
García-Quero, F., & Guardiola, J. (2018). Economic poverty and happiness in rural Ecuador: The importance of *Buen Vivir* (living well). *Applied Research in Quality of Life, 13*(4), 909–926. https://doi.org/10.1007/s11482-017-9566-z
Gore, T. (2015). Extreme carbon inequality: Why the Paris climate deal must put the poorest, lowest emitting and most vulnerable people first. Oxfam Media Briefing. https://oxfamilibrary.openrepository.com/handle/10546/582545
Guardiola, J., & García-Quero, F. (2014). Buen Vivir (living well) in Ecuador: Community and environmental satisfaction without household material prosperity? *Ecological Economics, 107*, 177–184. https://doi.org/10.1016/j.ecolecon.2014.07.032
Gudynas, E. (2011). Buen Vivir: Today's tomorrow. *Development, 54*(4), 441–447.
Guillén-Royo, M. (2016). *Sustainability and well-being*. Routledge.
Guillén-Royo, M., Guardiola, J., & Garcia-Quero, F. (2017). Sustainable development in times of economic crisis: A needs-based illustration from Granada (Spain). *Journal of Cleaner Production, 150*, 267–276.
Guillén-Royo, M., Velazco, J., & Camfield, L. (2013). Basic needs and wealth as independent determinants of happiness: An illustration from Thailand. *Social Indicators Research, 110*(2), 517–536. https://doi.org/10.1007/s11205-011-9941-3
Helne, T., & Hirvilammi, T. (2015). Wellbeing and sustainability: A relational approach. *Sustainable Development, 23*(3), 167–175. https://doi.org/10.1002/sd.1581
Hickel, J. (2017). Is global inequality getting better or worse? A critique of the World Bank's convergence narrative. *Third World Quarterly, 38*(10), 2208–2222. https://doi.org/10.1080/01436597.2017.1333414
Hidalgo-Capitán, A., & Cubillo-Guevara, A. P. (2014). Six open debates on Sumak Kawsay. *Iconos, 48*, 25–40.
Hirvilammi, T., & Helne, T. (2014). Changing paradigms: A sketch for sustainable wellbeing and ecosocial policy. *Sustainability, 6*(4), 2160–2175. https://doi.org/10.3390/su6042160
Kasser, T. (2002). *The high price of materialism*. The MIT Press.
Kasser, T. (2017). Living both well and sustainably: A review of the literature, with some reflections on future research, interventions and policy. *Philosophical Transactions of the Royal Society a: Mathematical, Physical and Engineering Sciences, 375*(2095), 20160369. https://doi.org/10.1098/rsta.2016.0369
López Castellano, F., García-Quero, F., & García-Carmona, M. (2018). Perspectives on human and social capital theories and the role of education: An approach from Mediterranean thought. *Educational Philosophy and Theory, 50*(14), 1421–1432.
López Martínez, M. (2017). *Noviolencia o barbarie*. Dyckinson.
Luechinger, S., & Raschky, P. A. (2009). Valuing flood disasters using the life satisfaction approach. *Journal of Public Economics, 93*(3–4), 620–633. https://doi.org/10.1016/j.jpubeco.2008.10.003
Meulewaeter, C. (2016). Los dividendos de la paz: Un estado de la cuestión. *Ámbitos. Revista De Estudios De Ciencias Sociales y Humanidades, 36*, 57–63.
Mies, M., & Shiva, V. (1993). *Ecofeminism*. Zed Books.
Milgram, S. (1965). Some conditions of obedience and disobedience to authority. *Human Relations, 18*, 57–66.
Naess, A. (2009). *The ecology of wisdom: Writings by Arne Naess*. Penguin.
Nisbet, E. K., Zelenski, J. M., & Murphy, S. A. (2011). Happiness is in our nature: Exploring nature relatedness as a contributor to subjective well-being. *Journal of Happiness Studies, 12*(2), 303–322. https://doi.org/10.1007/s10902-010-9197-7

O'Brien, C. (2008). Sustainable happiness: How happiness studies can contribute to a more sustainable future. *Canadian Psychology/psychologie Canadienne, 49*(4), 289–295. https://doi.org/10.1037/a0013235

O'Neill, D. W., Fanning, A. L., Lamb, W. F., & Steinberger, J. K. (2018). A good life for all within planetary boundaries. *Nature Sustainability, 1*(2), 88–95. https://doi.org/10.1038/s41893-018-0021-4

Perovic, L. M., & Golem, S. (2010). Investigating macroeconomic determinants of happiness in transition countries. *Eastern European Economics, 48*(4), 59–75.

Rojas, M., & Guardiola, J. (2017). Hunger and the experience of being well: Absolute and relative concerns. *World Development, 96*, 78–86.

Ruskin, J. (1985). *Unto this last.* Penguin UK.

Said, E. W. (1979). *Orientalism.* Vintage.

Salehyan, I. (2008). From climate change to conflict? No consensus yet. *Journal of Peace Research, 45*(3), 315–326. https://doi.org/10.1177/0022343308088812

Schraff, A. (2008). *Martin Luther King Jr.* Saddleback Educational Publishing.

Schumacher, E. F. (1973). *Small is beautiful: A study of economics as if people mattered.* Blond and Briggs. https://books.google.es/books/about/Small_Is_Beautiful.html?id=IKo3ALhVFKcC&redir_esc=y

Schwartz, S. H. (1992). Universals in the content and structure of values: Theoretical advances and empirical tests in 20 countries. *Advances in Experimental Social Psychology, 25*, 1–65.

Steinberger, J. K., & Roberts, J. T. (2010). From constraint to sufficiency: The decoupling of energy and carbon from human needs, 1975–2005. *Ecological Economics, 70*(2), 425–433. https://doi.org/10.1016/j.ecolecon.2010.09.014

Stiglitz, J., Sen, A., & Fitoussi, J. P. (2009). *Report by the Commission on the Measurement of Economic Performance and Social Progress* (Technical report). Commission on the Measurement of Economic Performance and Social Progress. https://ec.europa.eu/eurostat/documents/8131721/8131772/Stiglitz-Sen-Fitoussi-Commission-report.pdf

Suárez-Varela, M., Guardiola, J., & González-Gómez, F. (2016). Do pro-environmental behaviors and awareness contribute to improve subjective well-being? *Applied Research in Quality of Life, 11*(2), 429–444.https://doi.org/10.1007/s11482-014-9372-9

Tortosa, J. M. (2009). *Sumak Kawsay, Suma Qamaña, Buen Vivir*. Fundación Carolina.

Tortosa, J. M. (2012). Desigualdad, conflicto, violencia. Cinco ensayos sobre la realidad mundial. PYDLOS.

Urbina, D. A., & Ruiz-Villaverde, A. (2019). A Critical Review of Homo Economicus from Five Approaches. *American Journal of Economics and Sociology, 78*(1), 63–93.https://doi.org/10.1111/ajes.12258

Welsch, H. (2002). Preferences over prosperity and pollution: Environmental valuation based on happiness surveys. *Kyklos, 55*(4), 473–494. https://doi.org/10.1111/1467-6435.00198

World Values Survey (2019). *World values survey.* http://www.worldvaluessurvey.org/WVSOnline.jsp

Chapter 6
Where is the Double Dividend? The Relationship Between Different Types of Pro-environmental Behavior and Different Conceptions of Subjective Well-Being

Nazaret Ibáñez-Rueda and Jorge Guardiola Wanden-Berghe

Abstract The adoption of pro-environmental behaviors is essential to achieve the transition to a sustainable society. Previous evidence indicates that pro-environmental behaviors are positively associated with subjective well-being, lending support to the well-being dividend theory (namely that actions aimed at protecting the environment also improve well-being). In this study, we further examine the relationship between subjective well-being and pro-environmental behaviors, investigating whether the positive associations found extend to different types of behavior and different conceptions of well-being. We use three measures of well-being: life satisfaction (cognitive dimension), experienced emotions (affective dimension) and subjective vitality (eudaimonic dimension). We classify different pro-environmental behaviors on the basis of two criteria: the possible cost or benefit they entail for the individual who performs them, and the ease with which they can be observed by other people. Using regression analysis with data from a sample of students from the University of Granada, Spain, we found that the relationship between pro-environmental behaviors and subjective well-being differs according to the type of behavior and the dimension of well-being considered. Consistent with the well-being dividend theory, we found that actions that involve saving money are positively related to emotions, while actions that involve a cost, both in terms of money and time, are positively related to subjective vitality. However, we found no association between well-being and behaviors that do not entail any cost or benefit, and nonsignificant relations of certain behaviors with particular happiness measures. The nonsignificant relationships found between well-being and some categories of pro-environmental actions call for more

N. Ibáñez-Rueda (✉)
Department of Applied Economics, Universidad de Granada, Granada, Spain
e-mail: nibanez@ugr.es

J. G. Wanden-Berghe
Institute of Peace and Conflicts, Department of Applied Economics, Universidad de Granada, Granada, Spain
e-mail: jguardiola@ugr.es

S. Cloutier et al. (eds.), *Linking Sustainability and Happiness*,
Community Quality-of-Life and Well-Being,
https://doi.org/10.1007/978-3-030-89559-4_6

research and political action to link sustainable behavior with well-being, in order to simultaneously boost happiness and pro-environmental behavior.

Keywords Pro-environmental behavior · Subjective well-being · Well-being dividend

6.1 Introduction

The growing concern for the environment has fostered sustainability research in a variety of disciplines. In this regard, one way to contribute to the sustainability literature is through research on subjective well-being,[1] which can be key to encouraging sustainable behaviors,[2] such as recycling, reusing products or reducing overall consumption. The environmental situation requires us to change our behavior patterns, but caring for the environment does not have to be a sacrifice. Several studies reveal that people's well-being is positively correlated with how often they engage in pro-environmental behaviors, demonstrating that one can be happy caring for the environment (Binder & Blankenberg, 2017; Kaida & Kaida, 2016; Schmitt et al., 2018). Research on subjective well-being develops the concept of sustainable happiness and helps to discredit the dogma that consumption leads to happiness. In this vein, O'Brien (2008) defines sustainable happiness as the pursuit of happiness without exploiting other people, the environment or future generations. Similarly, Ulluwishewa (2016) points out that in order to achieve sustainability and happiness, we have to change the way we interact with other human beings and nature, abandoning selfish and exploitative relationships. Uniting the notions of sustainability and happiness creates potential for individual and global well-being.

Numerous studies find evidence of a positive relationship between ecological behavior and happiness (Zawadzki et al., 2020), helping to bolster the theory of the double dividend (Jackson, 2005), which argues that acting pro-environmentally is not only good for the environment, but is also accompanied by higher levels of happiness (Guillen-Royo, 2010; Guillen-Royo et al., 2017). However, the previous literature does not conclusively clarify whether this positive relationship occurs in a generalized way: it could be limited to behaviors with certain characteristics, or it may depend on the dimension of well-being considered. Furthermore, most of the previous studies approximate subjective well-being through life satisfaction, taking into account a single conceptualization of well-being.

In this paper, we build on the double dividend theory, seeking to further explore the relationship between pro-environmental behavior and subjective well-being, using

[1] We use subjective well-being (or just well-being) as a synonym for happiness to broadly refer to the experience of being well. This ranges from cognitive judgments and affective evaluations to being fully functioning (Diener et al., 2002; Ryan & Deci, 2001). Thus, we cover the cognitive, affective and eudaimonic dimensions of well-being.

[2] We use the terms *pro-environmental behavior*, *ecological behavior*, or *sustainable behavior* to refer to any responsible behavior that minimizes damage to, or even benefits, the environment (Steg & Vlek, 2009).

different conceptions of well-being and different categories of pro-environmental behavior. We consider three dimensions of well-being: the affective dimension, which is based on the PANAS scale (Watson et al., 1988); the cognitive dimension, approximated by life satisfaction (Dolan et al., 2011); and the eudaimonic dimension, which is reflected through subjective vitality (Guillen-Royo, 2019; Ryan & Frederick, 1997). We group the different pro-environmental behaviors into different categories: on the one hand, we classify these behaviors on the basis of whether they entail a monetary cost or saving, entail an extra cost in terms of time, or do not entail either a cost or a saving; on the other hand, we classify them according to whether or not they can be easily observed by third parties. To the best of our knowledge, this is the first time that different dimensions of well-being have been related to different categories of pro-environmental behavior, differentiating behaviors on the basis of their characteristics. The estimations are based on a sample of 973 students from the University of Granada, Spain.

The rest of this study is divided into the following sections: in Sect. 6.2 we present a description of the empirical evidence on the relationship between pro-environmental behavior and well-being; in Sect. 6.3 we describe the data used in the research, present the hypotheses and the method of analysis; in Sect. 6.4 we detail the results obtained; in Sect. 6.5 we present a discussion of the results and reflect on the limitations of this research and future research opportunities; and finally, in Sect. 6.6 we present our conclusion.

6.2 Literature Review: Well-Being and the Environment

The environment and well-being are closely related concepts, and a growing amount of research is focusing on the link between the two. There is increasing evidence indicating that behaving in a sustainable way also brings about greater levels of well-being (Zawadzki et al., 2020), supporting Jackson's (2005) conceptualization of the well-being dividend.

This theory, which comes with strong empirical evidence, calls into question the popular belief that pro environmental behaviors involve a sacrifice that diminishes quality of life. In fact, several studies indicate that frequent involvement in 2014 pro-environmental behaviors correlates positively with well-being (Kasser, 2017). In this respect, Suárez-Varela et al. (2014) find, for a sample from Granada, Spain, that awareness and pro-environmental actions aimed at saving water have a positive or non-significant influence on subjective well-being, but never a negative influence. Using samples from Canada and the United States, Schmitt et al. (2018) find that participation in 39 pro-environmental behaviors predicts greater life satisfaction in all cases except for using public transportation or carpooling, and using the washer or dryer only when they are full. Binder and Blankenberg (2016) study environmental activism in Germany, finding that environmental concerns lead to a greater propensity for volunteering and this volunteering is positively associated with well-being for those who are very concerned about the environment. For a sample from Stockholm,

Sweden, Kaida and Kaida (2016) find that pro-environmental behavior aimed at saving water and energy in the home not only improves current subjective well-being, but also improves expectations of future subjective well-being. Using data from Great Britain, Binder and Blankenberg (2017) replicate the positive relationship between life satisfaction and pro-environmental behavior.

In sum, research has demonstrated the positive relationship between pro-environmental behavior and subjective well-being, thus providing evidence to support the double dividend theory (Jackson, 2005). However, some studies found exceptions to the theory. The paper by Schmitt et al. (2018) reports that the relationship between well-being and pro-environmental behavior varies according to the characteristics of the behavior. In particular, the authors find that life satisfaction is most strongly predicted by behaviors that involve greater social interaction, behaviors that can be easily observed by others and behaviors that involve direct costs in terms of money, time and effort. This could explain why the double dividend may be more obvious in some research than in others. Yet, in addition to differences in the intensity of the association, a few studies show negative relationships between certain pro-environmental behaviors and well-being. For example, Ibáñez-Rueda et al. (2020) find, in a Spanish sample, that environmental activism is negatively associated with life satisfaction. Also using a sample from Spain, Binder et al. (2020) report a negative relationship between life satisfaction and a behavior index composed of 20 different environmentally-friendly actions. Similarly, Verhofstadt et al. (2016) point out that some ecological behaviors, such as reducing meat consumption or not using the car, are related to lower life satisfaction in a Belgian sample.

Taking into account these few exceptions to this theory, further exploration of the well-being dividend theory is merited. To that end, and in order to achieve the objective of this research, we test the relationship between different categories of pro-environmental actions and well-being.

6.3 Methodology

6.3.1 Field Work

In this research, we use a new database comprising 1283 students from the University of Granada, Spain. The field work was undertaken during the months of March and April 2019. A research team visited classrooms and provided a link to the questionnaire, which was accessible online via Qualtrics. The students did not receive any payment for filling in the questionnaire. After deleting missing values, as well as 5 observations that made no sense (they appeared to be answered randomly), the final sample used in our research comprised 973 observations.

6.3.2 Variables and Hypotheses

To evaluate the subjective well-being of individuals we use three indicators: life satisfaction, experienced affection and subjective vitality. Each one reflects a different understanding of subjective well-being.

Life satisfaction is related to the cognitive assessments and judgments people make about their life when they think about it (Dolan et al., 2008). We measure the variable *life satisfaction* through the question "How satisfied are you at this moment with your life as a whole?" Survey participants answered the question using an 11-point Likert scale ranging from 0 (*completely dissatisfied*) to 10 (*completely satisfied*).

Through the feelings and emotions an individual experiences, we evaluate the affective component of subjective well-being. We use the *Positive and Negative Affect Schedule* (PANAS) proposed by Watson et al. (1988), which is composed of 20 items describing different feelings and emotions, 10 relating to positive affect (*motivated, alert, excited, inspired, strong, determined, attentive, enthusiastic, active, proud*) and 10 to negative affect (*irritable, annoyed, upset, embarrassed, angry, nervous, guilty, fearful, aggressive, restless, insecure*). For each item, the individuals indicated how much they had felt these during the last seven days using a 5-point Likert scale (1 = very slightly or not at all, 2 = a little, 3 = moderately, 4 = quite a lot, and 5 = extremely). The variable *emotions* is calculated as the difference between the sum of the positive affect scores and the sum of the negative affect scores.

Finally, we use subjective vitality to reflect the eudaimonic dimension of well-being. Subjective vitality can be defined as the conscious experience of possessing energy and vivacity (Ryan & Frederick, 1997), and is considered an aspect of eudaimonic well-being because it is part of being in a state of full psychological and physical functioning (Guillen-Royo, 2019). We calculate the variable *vitality* as the arithmetic mean of the scores given to six statements related to a feeling of vitality: "I feel alive and vital"; "Sometimes I am so alive I just want to burst"; "I have positive energy and vigor"; "I get excited each new day"; "I am usually alert and awake"; "I feel full of energy." The answers to each statement are indicated on a 5-point scale (1 = Totally false; 2 = Not so true; 3 = Somewhat true; 4 = Fairly true; 5 = Extremely true).

We approximate pro-environmental behavior by asking individuals to score the frequency with which they perform the actions in Table 6.1. Participants indicated whether they engaged in these behaviors very little or not at all (1); a little (2); moderately (3); quite a lot (4); or extremely often (5). They were also given the option to indicate if any of the behaviors were not applicable to them. We re-coded the answers to the item "Throw away food" so that high scores denote environmentally-friendly behavior in all cases.

Based on these responses, we calculate different specific pro-environmental behavior indices using a taxonomy that incorporates the characteristics of each pro-environmental behavior. To this end, we establish two different classifications, based on the cost or benefit to the individual who performs the behavior and based on the ease of being observed by others. Thus, we first classify the actions into different

Table 6.1 Classification of pro-environmental behaviors

	According to cost				According to observability	
	Cost money	Cost time	Save money	Neutral (no cost or savings)	Easily observed	Not easily observed
Turn off lights in rooms that are not being used			X			X
Put on more clothes when it's cold at home, instead of turning on or turning up the heating			X			X
Decide not to buy something because it has an excess of packaging		X				X
Buy recycled products such as recycled toilet paper or tissues	X				X	
Take your own bag with you when you go shopping				X	X	
Separate the garbage. For example, paper, plastic, glass		X			X	
Use public transport (e.g. bus, train) instead of using the car			X		X	
Walk or cycle for short distances			X		X	
Avoid taking planes when possible			X			X
Participate in demonstrations in support of the environment		X			X	

(continued)

Table 6.1 (continued)

	According to cost				According to observability	
Reduce consumption of meat or animal products				X		X
Buy organic or eco-labelled food	X				X	
Buy organic or eco-labelled products (furniture, clothing)	X				X	
Preference for buying local products		X				X
Don't throw away food			X			X
In general, try to reduce consumption in everyday life			X		X	

categories according to whether they involve a monetary cost, a cost in terms of time, a saving of money, or do not involve a cost or a saving. Secondly, we distinguish between behaviors that are easily observed by others and behaviors that are more private and difficult for third parties to observe. Table 6.1 shows this classification of behaviors. Each index is calculated as the average of the scores obtained in the behaviors of the specific category. To calculate this average, the sum of the scores is divided by the number of behaviors in that particular category to which the individual responded, so that the index score is not affected if one of the behaviors is not applicable to the individual.

Based on the literature reviewed in Sect. 6.2, we expect that, in general, the most frequent participation in pro-environmental behaviors is positively related to subjective well-being (Hypothesis 1), thus confirming a double dividend. This hypothesis has been widely tested in the literature, however, most studies that analyze this relationship tend to consider only the cognitive dimension of well-being. We expect that this positive association also occurs for the affective and eudaimonic dimensions.

Secondly, we want to investigate whether the intensity of the relationship between pro-environmental behavior and subjective well-being differs according to the type of behavior (Hypothesis 2). This hypothesis is strongly supported by the study of Schmitt et al. (2018). The findings of Ibáñez-Rueda et al. (2020) are also consistent with this hypothesis, as they find that sustainable consumption and environmental activism have different associations with subjective well-being.

Finally, we explore the hypothesis that the magnitude of the relationship between different types of pro-environmental behavior and well-being differs according to the dimension of subjective well-being considered (Hypothesis 3). This hypothesis complements hypothesis 2 and constitutes the main innovation of our research because, although some previous studies consider different types of behavior and different measures of well-being, their contributions are different from those of this research. For example, Martin et al. (2020) include two indicators of pro-environmental behavior (*household* and *nature conservation*) and two measures of subjective well-being (evaluative and eudaimonic well-being), but treat both concepts as outcome variables, without analyzing the relationship between them. Although Ibáñez-Rueda et al. (2020) do analyze the relationship between different types of pro-environmental behavior (sustainable consumption and environmental activism) and different measures of subjective well-being (cognitive, affective and eudaimonic), the distinction between pro-environmental behavior is based on conceptual criteria and not on the specific characteristics of the different activities. Therefore, this hypothesis is not entirely supported by previous studies, even though some research shows that different well-being dimensions correlate differently with different observable characteristics (Graham & Nikolova, 2015; Rojas & Guardiola, 2017).

In order to test these hypotheses, in addition to the main variables of well-being and pro-environmental behavior, we include a set of control variables capturing respondents' personal information.

Income. The participants indicated their parents' monthly income by selecting one of the eight intervals given as an option, with the lowest category being less than €499 and the highest €5000 or more. We estimated the income for each category using the midpoint of the interval (except in the case of the top category, where we estimated it at €6000). We calculate per capita income dividing by the number of people living in the household. In the analysis we include the natural logarithm of these incomes.

Age. We consider the age in years specified by respondents. We also incorporate the variable age^2.

Gender. Respondents indicated their gender by selecting male (1); female (2); or other (3).

Marital status. This variable comprises five categories: married (1); with a stable partner (2); divorced or separated (3); widowed (4); and single (5).

Relationships. The variable *relationships* reflects respondents' social life. They are asked about the frequency with which they are in touch with their relatives, friends and neighbors on a scale of 1 (never) to 5 (every day or almost every day). We calculate the variable as the average of the scores obtained in the three items (family, friends and neighbors).

Health. Individuals describe their health status by selecting one of the following categories: no problems (1); mild problems (2); moderate problems (3); and major problems (4).

Work. A dichotomous variable that indicates whether the respondent works (1) or not (0).

Area. This variable takes three values, depending on whether the participants live in a rural area/village (1); in an urban area, near a city (2); or in a city (3).

6.3.3 Method of Analysis

To examine the relationship between pro-environmental behavior and well-being, we use a regression analysis, which allows us to determine both the nature and strength of the relationship between these two variables. We specify different models for each dimension of well-being, incorporating life satisfaction, emotions, and subjective vitality as dependent variables, in turn. For each model, we perform a series of regression analyses, including the proposed categories of behavior, one by one, as the predictor variable.

For the estimation of the parameters, we use Ordinary Least Squares (OLS). This is the most suitable method for the variables *emotions* and *vitality*, since they can be treated as cardinal variables. In contrast, given the ordinal nature of the variable *life satisfaction*, it would be more appropriate to use an ordered probit model. However, as in previous studies (e.g., Guillen-Royo, 2019; Verhofstadt et al., 2016), we apply OLS because its interpretation is simpler and the results obtained with these two methods are very similar (Ferrer-i-Carbonell & Frijters, 2004). In any case, we repeat the analyses referring to life satisfaction using ordered probit models, finding similar results. Data analysis was performed using Stata15 statistical software.

6.4 Results

First, we performed a descriptive analysis of the variables used in the regression analysis. Table 6.2 presents the descriptive statistics of these variables. With regard to ecological behavior, distinguishing according to the cost/benefit they involve, those that represent savings register the highest frequency, while those that represent a monetary cost are the least common. On the other hand, actions that are not easily observed by others are more common than observable actions.

Next, we studied the relationship of the different types of actions with life satisfaction (Table 6.3), emotions experienced (Table 6.4) and vitality (Table 6.5). For each conception of well-being, we first analyzed the relationship with pro-environmental behaviors differentiated according to the cost/benefit for the individual who performs them (*a*). Secondly, we studied this relationship by distinguishing between behaviors according to the ease with which they are observed by third parties (*b*).

While some types of behavior are positively and significantly associated with the affective or eudaimonic dimensions of subjective well-being, the category of neutral

Table 6.2 Descriptive statistics

	Mean/%[a]	Std. Dev.	Min	Max
Subjective well-being				
Life satisfaction	7.0319	1.6985	1	10
Emotions	7.9928	11.0438	−25	35
Vitality	3.3097	0.7389	1	5
Pro-environmental behavior (PEB)				
PEB cost-money	1.9399	0.9128	1	5
PEB cost-time	2.4657	0.8855	1	5
PEB saving	3.6739	0.6160	1.1429	5
PEB neutral	2.9049	1.0983	1	5
PEB observable	2.8223	0.7161	1	5
PEB non-observable	3.1014	0.6208	1	5
Control variables				
Income	6.1916	0.7967	3.2189	8.6995
Age	20.7040	2.8121	18	54
Gender (female)	62.28%		0	1
Marital status (single)	63.21%		0	1
Relationships	3.4813	0.7542	1	5
Work	25.49%		0	1
Area (rural)	16.03%		0	1
Area (near a city)	16.96%		0	1
Area (city)	67.01%		0	1
Health (no problems)	53.55%		0	1
Health (mild problems)	37.31%		0	1
Health (moderate problems)	8.32%		0	1
Health (major problems)	0.82%		0	1

[a]This column shows the mean of the quantitative variables, while for the qualitative variables it shows the percentage of individuals in the sample with that particular characteristic

Table 6.3 Life satisfaction and different types of pro-environmental behavior

	Life satisfaction					
	(a)				*(b)*	
PEB cost-money	−0.019 (0.062)					
PEB cost-time		−0.0724 (0.065)				

(continued)

Table 6.3 (continued)

	Life satisfaction					
	(a)				*(b)*	
PEB saving			−0.0452 (0.096)			
PEB neutral				−0.0435 (0.047)		
PEB observable					−0.0517 (0.079)	
PEB non-observable						−0.102 (0.091)
Income	0.0978 (0.064)	0.0971 (0.064)	0.0952 (0.064)	0.095 (0.064)	0.0962 (0.064)	0.0931 (0.064)
Age	−0.0235 (0.064)	−0.0206 (0.064)	−0.0251 (0.065)	−0.0213 (0.064)	−0.0251 (0.065)	−0.0206 (0.064)
Age^2	0.000592 (0.00103)	0.000555 (0.00102)	0.000604 (0.00103)	0.000557 (0.00102)	0.000623 (0.00103)	0.000538 (0.00102)
Gender (female)	0.0754 (0.109)	0.0855 (0.109)	0.0796 (0.108)	0.0912 (0.110)	0.0819 (0.108)	0.0885 (0.108)
Marital status (single)	−0.454*** (0.106)	−0.457*** (0.106)	−0.454*** (0.106)	−0.460*** (0.106)	−0.456*** (0.106)	−0.457*** (0.106)
Relationships	0.360*** (0.072)	0.368*** (0.073)	0.358*** (0.072)	0.356*** (0.072)	0.363*** (0.073)	0.358*** (0.073)
Work	−0.0766 (0.128)	−0.0629 (0.129)	−0.0808 (0.128)	−0.0734 (0.128)	−0.0755 (0.128)	−0.0665 (0.128)
Area (near a city)	0.0103 (0.175)	0.00323 (0.175)	0.0117 (0.175)	0.00833 (0.175)	0.0108 (0.175)	0.00545 (0.175)
Area (city)	−0.0268 (0.144)	−0.0259 (0.144)	−0.0199 (0.145)	−0.0203 (0.144)	−0.0196 (0.145)	−0.0231 (0.144)
Health (mild problems)	−0.585*** (0.107)	−0.580*** (0.107)	−0.583*** (0.107)	−0.582*** (0.107)	−0.583*** (0.107)	−0.580*** (0.107)
Health (moderate problems)	−1.344*** (0.220)	−1.336*** (0.219)	−1.347*** (0.219)	−1.335*** (0.219)	−1.341*** (0.219)	−1.342*** (0.219)
Health (major problems)	−1.760*** (0.617)	−1.746*** (0.596)	−1.764*** (0.615)	−1.769*** (0.613)	−1.764*** (0.609)	−1.754*** (0.610)
Constant	6.059*** (1.047)	6.121*** (1.041)	6.231*** (1.153)	6.135*** (1.038)	6.179*** (1.073)	6.325*** (1.089)
F	9.84	9.9	9.87	9.86	9.85	9.92
R^2	0.12	0.121	0.12	0.12	0.12	0.121
N	973	973	973	973	973	973

Robust standard errors in parentheses
***$p < 0.01$, **$p < 0.05$, *$p < 0.1$

Table 6.4 Emotions index and different types of pro-environmental behavior

	Emotions					
	(a)				*(b)*	
PEB cost-money	0.305 (0.391)					
PEB cost-time		0.344 (0.386)				
PEB saving			0.941* (0.570)			
PEB neutral				0.272 (0.320)		
PEB observable					0.754 (0.506)	
PEB non-observable						0.427 (0.562)
Income	0.957** (0.432)	0.959** (0.431)	1.014** (0.430)	0.974** (0.431)	0.980** (0.430)	0.976** (0.431)
Age	0.0162 (0.439)	−0.00392 (0.438)	0.0521 (0.437)	−0.00307 (0.439)	0.039 (0.438)	−0.00252 (0.439)
Age^2	0.00448 (0.007)	0.00488 (0.007)	0.00413 (0.007)	0.0049 (0.007)	0.00406 (0.007)	0.00495 (0.007)
Gender (female)	−1.631** (0.711)	−1.676** (0.709)	−1.720** (0.709)	−1.728** (0.715)	−1.725** (0.710)	−1.683** (0.709)
Marital status (single)	−1.323* (0.717)	−1.315* (0.718)	−1.322* (0.714)	−1.294* (0.721)	−1.291* (0.717)	−1.318* (0.718)
Relationships	2.932*** (0.470)	2.908*** (0.469)	2.956*** (0.466)	2.968*** (0.468)	2.893*** (0.469)	2.957*** (0.467)
Work	0.13 (0.856)	0.106 (0.854)	0.2 (0.848)	0.146 (0.850)	0.119 (0.850)	0.131 (0.853)
Area (near a city)	−0.702 (1.228)	−0.67 (1.230)	−0.73 (1.227)	−0.691 (1.228)	−0.709 (1.227)	−0.684 (1.229)
Area (city)	0.41 (0.916)	0.416 (0.915)	0.26 (0.925)	0.378 (0.921)	0.306 (0.921)	0.405 (0.917)
Health (mild problems)	−3.792*** (0.716)	−3.817*** (0.716)	−3.829*** (0.715)	−3.811*** (0.715)	−3.819*** (0.716)	−3.815*** (0.716)
Health (moderate problems)	−5.772*** (1.363)	−5.777*** (1.356)	−5.734*** (1.359)	−5.802*** (1.359)	−5.824*** (1.358)	−5.746*** (1.357)
Health (major problems)	−12.98*** (3.583)	−13.04*** (3.643)	−12.89*** (3.663)	−12.92*** (3.638)	−12.92*** (3.703)	−13.00*** (3.604)
Constant	−7.357 (6.919)	−7.271 (6.855)	−11.09 (7.115)	−7.501 (6.846)	−9.064 (6.962)	−8.075 (6.907)
F	9.75	9.73	10.14	9.73	9.89	9.8

(continued)

Table 6.4 (continued)

	Emotions					
	(a)				*(b)*	
R^2	0.115	0.115	0.117	0.115	0.117	0.115
N	973	973	973	973	973	973

Robust standard errors in parentheses
***p < 0.01, **p < 0.05, *p < 0.1

Table 6.5 Vitality and different types of pro-environmental behavior

	Vitality					
	(a)				*(b)*	
PEB cost-money	0.0536** (0.0269)					
PEB cost-time		0.0611** (0.0256)				
PEB saving			0.0616 (0.0395)			
PEB neutral				0.0162 (0.0207)		
PEB observable					0.0666** (0.0321)	
PEB non-observable						0.0759** (0.0368)
Income	0.0541* (0.0278)	0.0546** (0.0278)	0.0577** (0.0278)	0.0550** (0.0279)	0.0561** (0.0278)	0.0575** (0.0277)
Age	0.0551** (0.0275)	0.0516* (0.0276)	0.0565** (0.0276)	0.0529* (0.0276)	0.0563** (0.0276)	0.0518* (0.0275)
Age^2	−0.00075* (0.0004)	−0.000675 (0.0004)	−0.00073* (0.0004)	−0.00068 (0.0004)	−0.00076* (0.0004)	−0.00066 (0.0004)
Gender (female)	−0.0507 (0.0469)	−0.0588 (0.0467)	−0.0561 (0.0468)	−0.0561 (0.0474)	−0.0588 (0.0469)	−0.06 (0.0466)
Marital status (single)	−0.0276 (0.0456)	−0.0262 (0.0456)	−0.029 (0.0456)	−0.0274 (0.0459)	−0.0259 (0.0456)	−0.0267 (0.0456)
Relationships	0.293*** (0.0322)	0.289*** (0.0318)	0.297*** (0.0318)	0.298*** (0.0319)	0.292*** (0.0321)	0.298*** (0.0317)
Work	0.0172 (0.0504)	0.0127 (0.0508)	0.0282 (0.0504)	0.0249 (0.0501)	0.0213 (0.0505)	0.0172 (0.0504)
Area (near a city)	−0.111 (0.0767)	−0.106 (0.0766)	−0.113 (0.0768)	−0.111 (0.077)	−0.112 (0.0767)	−0.108 (0.0767)

(continued)

Table 6.5 (continued)

	Vitality					
	(a)				*(b)*	
Area (city)	−0.0518 (0.0618)	−0.0507 (0.0617)	−0.0599 (0.0633)	−0.052 (0.0623)	−0.0596 (0.0627)	−0.0526 (0.0619)
Health (mild problems)	−0.226*** (0.0472)	−0.231*** (0.0473)	−0.229*** (0.0474)	−0.228*** (0.0474)	−0.229*** (0.0473)	−0.231*** (0.0474)
Health (moderate problems)	−0.417*** (0.0838)	−0.418*** (0.0838)	−0.409*** (0.0842)	−0.413*** (0.0833)	−0.417*** (0.0844)	−0.412*** (0.0835)
Health (major problems)	−0.682*** (0.184)	−0.693*** (0.188)	−0.676*** (0.184)	−0.678*** (0.185)	−0.676*** (0.184)	−0.686*** (0.189)
Constant	1.258*** (0.444)	1.272*** (0.439)	1.073** (0.467)	1.312*** (0.437)	1.154*** (0.446)	1.129** (0.447)
F	15.51	15.33	15.06	14.98	15.31	15.11
R^2	0.164	0.165	0.162	0.16	0.164	0.164
N	973	973	973	973	973	973

Robust standard errors in parentheses
***$p < 0.01$, **$p < 0.05$, *$p < 0.1$

behavior (actions that do not involve either a cost or a saving) is not associated with any dimension of well-being. In contrast, no behavior is significantly related to life satisfaction. The emotions experienced are related only to the behaviors that entail a saving ($b = 0.941$, $p < 0.1$). Subjective vitality is associated with behaviors that involve a cost, in terms of money ($b = 0.0536$, $p < 0.05$) and time ($b = 0.0613$, $p < 0.01$). Furthermore, it is related both to behaviors that are easily observed by others ($b = 0.0666$, $p < 0.05$), and to those that are not ($b = 0.0759$, $p < 0.05$). Therefore, we do not find differences between the type of behavior based on this last criterion. The rest of the relationships between subjective well-being and behaviors are nonsignificant.

In summary, the results obtained only partially support the first hypothesis formulated above: in general, more frequent participation in pro-environmental behavior is positively related only to the affective and eudaimonic dimensions of subjective well-being, whereas it is not significant for the cognitive dimension. The findings are consistent with hypotheses 2 and 3: the relationship between pro-environmental behavior and well-being differs according to the type of behavior and the dimension of well-being considered.

6.5 Discussion

Several studies have concluded that caring for the environment and improving people's well-being are compatible goals. A growing body of empirical evidence on this has strengthened the double dividend theory: being environmentally responsible results in a benefit both for the environment and for people's well-being (Jackson, 2005). However, the empirical evidence in this research does not universally reproduce the existence of the double dividend. While the theory may hold true for the scenario under study, the challenge arises of extending the double dividend to all scenarios. The results of this study underline the importance of the concept of well-being and the type of behavior taken into account, and call for further research on different specific scenarios, which could shape policies aiming to create a double dividend.

As an analysis strategy, we applied two different categorizations of pro-environmental actions: firstly, differentiating according to the cost/benefit for the individual who performs them; and secondly, according to the ease with which these behaviors are observed by third parties. In the first part of the analysis, we find that saving behaviors are positively related to the emotions experienced, while cost behaviors are positively associated with subjective vitality. The behaviors that do not entail either a cost or savings are not related to any dimension of well-being. In the second part of the analysis, which considers whether or not behaviors are observed by third parties, we find no differences, as both types are related only to subjective vitality. The fact that a behavior can be easily observed by others does not seem to have any effect on the relationship between sustainable behavior and well-being.

These results have two implications for the theory of the well-being dividend that merit discussion. Firstly, the evidence that saving behaviors are positively related to the affective dimension of well-being supports the theory. In this regard, this evidence is in accordance with some past research, with particular differences regarding the well-being dimension under consideration. Previous studies have found a positive relationship between saving behaviors and happiness (Gokdemir, 2015; Guven, 2012). The results concerning costly behaviors are in line with those obtained by Schmitt et al. (2018), in that it seems that such behaviors have a stronger (positive) relationship with well-being than those that do not involve cost. However, the associations they found were with life satisfaction, while ours are for subjective vitality. More costly behaviors may be more strongly related to well-being because they may be perceived as more effective in helping the environment (Schmitt et al., 2018).

The second implication with respect to this theory refers to specific nonsignificant relationships we have found between well-being and sustainable behavior. We found no significant positive associations between life satisfaction and pro-environmental behavior, unlike several previous studies (Brown & Kasser, 2005; Kaida & Kaida, 2016; Xiao & Li, 2011). However, this finding is not entirely new in the literature. Some previous research has already found nonsignificant and even negative relationships with life satisfaction (Binder et al., 2020; Ibáñez-Rueda et al., 2020; Suarez-Varela et al., 2014; Verhofstadt et al., 2016). In this respect, a possible explanation can be drawn from the role of pro-environmental values as a social norm. Welsch and Kühling (2018) found that part of the well-being benefit of holding pro-environmental values derives from compliance with a shared social norm. Perhaps greater unanimity in pro-environmental attitudes is needed to benefit from increased life satisfaction through environmentally-responsible behavior. Furthermore, the possibility that individuals do not see sustainable behaviors as a well-established social norm may also explain why no differences have been found between behaviors that are easily observed by others and those that are not. However, this is left as a hypothesis for further research, as we could not test it in the present study.

In our view, we do not think that our evidence invalidates the existence of the well-being dividend. Rather, it raises the possibility that it may be not present in certain groups. Further research would shed light on whether it can be reinforced through political proposals such as educational programs. Future research should also consider different social groups and classes, different cultures, the presence of social norms, the specific characteristics of individuals, behavioral characteristics, or different well-being conceptions. The last two factors seem relevant in view of our results, although they merit further exploration, and alternative factors should be considered. To this end, it would be advisable to reproduce this study with samples from other countries and covering a more diverse population, not limited to students. The latter aspect is especially interesting for the existing evidence. As Binder et al. (2020) argue, students are generally characterized by low income and some studies have pointed out that disposable income could affect the performance of certain pro-environmental actions (Clark et al., 2003; Stern et al., 1999; Zorić & Hrovatin, 2012). In addition, it might also be interesting to consider a wider selection of pro-environmental actions.

Finally, it should be noted that one of the main limitations of the results obtained is that they can only be interpreted as an association, and thus do not allow us to establish a cause-effect relationship. Therefore, the relationship between the two concepts studied may go in either direction.

6.6 Conclusion

In this research, we have examined the relationship between pro-environmental behavior and subjective well-being, exploring whether the positive associations found by previous research extend to different types of behavior and conceptions of well-being. Using regression analysis with data from a sample of students from the University of Granada, Spain, we found that the relationship between pro-environmental behavior and well-being differs according to the type of behavior and the dimension of well-being considered. More frequent participation in pro-environmental behavior is positively related only to the affective (experienced emotions) and eudaimonic (subjective vitality) dimensions of well-being. We did not find significant associations between any type of pro-environmental behavior and life satisfaction. In relation to the cost/benefits of the behaviors, we found that actions that involve saving money are positively related to emotions, while actions that involve a cost, in terms of money and time, are positively related to subjective vitality. In contrast, we found no association with well-being for behaviors that entail no cost or benefit to the individual who performs them. As regards their relationship with subjective well-being, we did not find differences between behaviors that are easily observable by third parties and those that are not, with both being positively related to subjective vitality. Despite some exceptions, the well-being dividend theory is still supported by empirical results from numerous studies. However, our particular results raise the possibility that some political actions are needed to link happiness and sustainability in particular contexts. Future research is faced with the task of identifying those specific scenarios and finding out why the theory does not hold there.

Acknowledgements The authors acknowledge the financial support from the Spanish Agencia Estatal de Investigación and the European Regional Development Fund (Project ECO2017-86822-R); the Regional Government of Andalusia and the European Regional Development Fund (Projects P18-RT-576 and B-SEJ-018-UGR18) and the University of Granada (Plan Propio. Unidad Científica de Excelencia: Desigualdad, Derechos Humanos y Sostenibilidad -DEHUSO-).

References

Binder, M., & Blankenberg, A. K. (2016). Environmental concerns, volunteering and subjective well-being: Antecedents and outcomes of environmental activism in Germany. *Ecological Economics, 124*, 1–16.

Binder, M., & Blankenberg, A. K. (2017). Green lifestyles and subjective well-being: More about self-image than actual behavior? *Journal of Economic Behavior & Organization, 137*, 304–323.

Binder, M., Blankenberg, A. K., & Guardiola, J. (2020). Does it have to be a sacrifice? Different notions of the good life, pro-environmental behavior and their heterogeneous impact on well-being. *Ecological Economics, 167*, 106448.

Brown, K. W., & Kasser, T. (2005). Are psychological and ecological well-being compatible? The role of values, mindfulness, and lifestyle. *Social Indicators Research, 74*(2), 349–368.

Clark, C. F., Kotchen, M. J., & Moore, M. R. (2003). Internal and external influences on pro-environmental behavior: Participation in a green electricity program. *Journal of Environmental Psychology, 23*(3), 237–246.

Diener, E., Lucas, R. E., & Oishi, S. (2002). Subjective well-being: The science of happiness and life satisfaction. In C. R. Snyder & S. J. Lopez (Eds.), *Handbook of positive psychology* (pp. 63–73). Oxford University Press.

Dolan, P., Layard, R., & Metcalfe, R. (2011). *Measuring subjective wellbeing for public policy: Recommendations on measures (No. 23)*. Centre for Economic Performance, LSE.

Dolan, P., Peasgood, T., & White, M. (2008). Do we really know what makes us happy? A review of the economic literature on the factors associated with subjective well-being. *Journal of Economic Psychology, 29*(1), 94–122.

Ferrer-i-Carbonell, A., & Frijters, P. (2004). How important is methodology for the estimates of the determinants of happiness? *The Economic Journal, 114*, 641–659.

Gokdemir, O. (2015). Consumption, savings and life satisfaction: The Turkish case. *International Review of Economics, 62*, 183–196.

Graham, C., & Nikolova, M. (2015). Bentham or Aristotle in the development process? An empirical investigation of capabilities and subjective well-being. *World Development, 68*(1), 163–179.

Guillen-Royo, M. (2010). Realising the "wellbeing dividend": An exploratory study using the Human Scale Development approach. *Ecological Economics, 70*(2), 384–393.

Guillen-Royo, M. (2019). Sustainable consumption and wellbeing: Does on-line shopping matter? *Journal of Cleaner Production, 229*, 1112–1124.

Guillen-Royo, M., Guardiola, J., & Garcia-Quero, F. (2017). Sustainable development in times of economic crisis: A needs-based illustration from Granada (Spain). *Journal of Cleaner Production, 150*, 267–276.

Guven, C. (2012). Reversing the question: Does happiness affect consumption and savings behavior? *Journal of Economic Psychology, 33*(4), 701–717.

Ibáñez-Rueda, N., Guillén-Royo, M., & Guardiola, J. (2020). Pro-environmental behavior, connectedness to nature, and wellbeing dimensions among Granada students. *Sustainability, 12*(21), 9171.

Jackson, T. (2005). Live better by consuming less? Is there a double dividend in sustainable consumption? *Journal of Industrial Ecology, 9*(1–2), 19–36.

Kaida, N., & Kaida, K. (2016). Pro-environmental behavior correlates with present and future subjective well-being. *Environment, Development and Sustainability, 18*(1), 111–127.

Kasser, T. (2017). Living both well and sustainably: A review of the literature, with some reflections on future research, interventions and policy. *Philosophical Transactions of the Royal Society a: Mathematical, Physical and Engineering Sciences, 375*(2095), 20160369.

Martin, L., White, M., Hunt, A., Richardson, M., Pahl, S., & Burt, J. (2020). Nature contact, nature connectedness and associations with health, wellbeing and pro-environmental behaviours. *Journal of Environmental Psychology, 68*, 101389.

O'Brien, C. (2008). Sustainable happiness: How happiness studies can contribute to a more sustainable future. *Canadian Psychology/psychologie Canadienne, 49*(4), 289–295. https://doi.org/10.1037/a0013235

Rojas, M., & Guardiola, J. (2017). Hunger and the experience of being well: Absolute and relative concerns. *World Development, 96*, 78–86.

Ryan, R. M., & Deci, E. L. (2001). On happiness and human potentials: A review of research on hedonic and eudaimonic well-being. *Annual Review of Psychology, 52*(1), 141–166.

Ryan, R. M., & Frederick, C. (1997). On energy, personality, and health: Subjective vitality as a dynamic reflection of well-being. *Journal of Personality, 65*(3), 529–565.

Schmitt, M. T., Aknin, L. B., Axsen, J., & Shwom, R. L. (2018). Unpacking the relationships between pro-environmental behavior, life satisfaction, and perceived ecological threat. *Ecological Economics, 143*, 130–140.

Steg, L., & Vlek, C. (2009). Encouraging pro-environmental behaviour: An integrative review and research agenda. *Journal of Environmental Psychology, 29*(3), 309–317.

Stern, P. C., Dietz, T., Abel, T., Guagnano, G. A., & Kalof, L. (1999). A value-belief norm theory of support for social movements: The case of environmentalism. *Human Ecology Review, 6*(2), 81–97.

Suárez-Varela, M., Guardiola, J., & González-Gómez, F. (2014). Do pro-environmental behaviors and awareness contribute to improve subjective well-being? *Applied Research in Quality of Life, 11*(2), 429–444. https://doi.org/10.1007/s11482-014-9372-9

Ulluwishewa, R. (2016). Spirituality, sustainability and happiness: A Quantum-neuroscientific perspective. In S. Dhiman & J. Marques (Eds.), *Spirituality and sustainability: New horizons and exemplary approach* (pp. 155–168). Springer.

Verhofstadt, E., Van Ootegem, L., Defloor, B., & Bleys, B. (2016). Linking individuals' ecological footprint to their subjective well-being. *Ecological Economics, 127*, 80–89.

Watson, D., Clark, L. A., & Tellegen, A. (1988). Development and validation of brief measures of positive and negative affect: The PANAS scales. *Journal of Personality and Social Psychology, 54*(6), 1063–1070.

Welsch, H., & Kühling, J. (2018). How green self image is related to subjective well-being: Pro-environmental values as a social norm. *Ecological Economics, 149*, 105–119.

Xiao, J. J., & Li, H. (2011). Sustainable consumption and life satisfaction. *Social Indicators Research, 104*(2), 323–329.

Zawadzki, S. J., Steg, L., & Bouman, T. (2020). Meta-analytic evidence for a robust and positive association between individuals' pro-environmental behaviors and their subjective wellbeing. *Environmental Research Letters, 15,* 123007.

Zorić, J., & Hrovatin, N. (2012). Household willingness to pay for green electricity in Slovenia. *Energy Policy, 47*, 180–187.

Chapter 7
Are the Sustainable Development Goals the Compass for a Happier Society?

Leire Iriarte

Abstract The Sustainable Development Goals (SDGs), agreed upon in 2015, are a cornerstone of decades of efforts toward more sustainable development. The SDGs include a holistic view of development including the social, economic and environmental dimensions of development and an underlying assumption that they contribute to happiness and well-being but still there are many unknown relations and effects that need analysis and reflection. This chapter explores how the SDGs and various environmental indicators interact with the happiness and well-being measures at the country level and reflect on some of the implications. We found a significant and positive correlation level between the Subjective Wellbeing levels, SDGs index, and various Environmental Indicators at the country level. Thus, the higher the Subjective Wellbeing, the higher the SDGs index and the higher the environmental impact (in terms of ecological footprint and spillover effect). Next, we discuss the intrinsic synergies and trade-offs within the environmental, social and economic dimensions of the SDGs and the need to decouple well-being levels from consumption levels and their respective environmental impacts. Learning from previous sustainable development experiences, we propose to promote more conscious societies by means of encouraging intrinsic values and motivations as well as mindfulness.

Keywords Sustainable development goals (SDGs) · Happiness and well-being metrics · Ecological footprint · Well-being

7.1 Introduction

In the 1987 Brundtland Report, the concept of sustainable development was defined as, "the development that meets the needs of the present without compromising the ability of future generations to meet their own needs" (Keeble, 1988). The purpose of sustainable development is to improve the quality of life of people while respecting the environment, also a cornerstone concept in international debates.

L. Iriarte (✉)
Founder and Director, El Buen Vivir, Navarra, Spain
e-mail: hola@elbuenvivir.org

S. Cloutier et al. (eds.), *Linking Sustainability and Happiness*,
Community Quality-of-Life and Well-Being,
https://doi.org/10.1007/978-3-030-89559-4_7

The General Assembly of United Nations (UN) agreed upon the SDGs in 2015 (United Nations, 2015) after long negotiations and previous international development processes such as the Rio Summit in 1992 or the Millennium Development Goals. In comparison with previous development processes, the SDGs were more ambitious in different ways: their relevance for all countries regardless of their classification as developed and developing states; their multisectoral composition considering goals related to people, planet, prosperity, peace and partnership and the importance of being a collaborative effort in which the governments, the private sector, academia and civil society are called to contribute. The SDGs are aspirational and the 193 signatory countries have the right to decide how far their aspirations should go. Given the interdependency among the 17 goals, synergies or trade-offs can be created (Nilsson et al., 2016). A comprehensive assessment of the progress by country towards the SDGs is provided by the Sustainable Development Solutions Network (Sachs et al., 2020).

Happiness and well-being are the result of many social, environmental and economic factors including freedom, equality, good social relationships and many others. According to the World Happiness Report (Helliwell et al., 2020), the average life satisfaction in a country is explained by six factors, including: GDP per capita, social support, healthy life expectancy at birth, freedom to make life choices, generosity and perceptions of corruption. Our happiness and well-being levels have multiple effects on a personal level in relation, for example, to our attitudes toward sustainability. The economic, social, and health context of a nation is also relevant for the levels of human well-being (De Neve et al., 2013). Over the past decade a steady decline of self-assessed happiness levels has been observed and, to date, the average effect of the COVID-19 pandemic on individual happiness ratings at the global level has not substantially changed (IPSOS, 2020).

Despite the importance of well-being and happiness, the globally accepted indicator to measure progress has been the Gross Domestic Product (GDP) which is a purely econometric indicator. However, recently, in response to the limitations of GDP as indicator of progress as discussed in numerous studies (Stiglitz et al., 2019), experts of economic, social, psychological and environmental fields have proposed many alternative indicators to GDP including those related to the happiness and well-being family. One of the reasons why happiness and well-being indices might be of interest is because many different types of well-being metrics support better policy making by providing feedback on whether long-term high-level policy objectives are being achieved (Durand & Exton, 2019).

Currently, there are many different proposals of measures for happiness and well-being at various governance levels, with different scopes, associated methods and various leading institutions. Among others, the Organization for Economic Co-operation and Development (OECD) developed the Better Life Index. Bhutan has been applying the Gross National Happiness Index (GNHI) for decades and the UK has developed its own measures on well-being. These measures vary in the number of indicators, the topics addressed and the number and nature of subjective measurements considered. Despite the various efforts and contrary to what occurs with the

SDGs, there is not yet an international agreed measure for happiness and well-being, and, perhaps, there should not be.

Despite the utmost importance of the concepts of and relations between sustainability and happiness, fostering happiness within a community is not currently a goal of mainstream sustainable community development frameworks (Cloutier & Pfeiffer, 2015). Despite this general lack of bridging between terms, in certain proposals the well-being lenses have started to be considered to approach, for example, climate change, as in the OECD report *Accelerating climate action refocusing policies through a well-being lens* (OECD, 2019).

The SDGs and the happiness and well-being indices share many overarching topics such as health and education. In fact, the OECD (2018) compared its well-being framework with the SDGs and concluded that the two approaches are complementary. The OECD well-being framework is an analytic and diagnostic tool to assess the conditions of a community, whereas the 2030 Agenda (United Nations, 2015) is a list of policy commitments agreed upon by world leaders. The SDGs touch on practically all of the dimensions considered in the OECD well-being framework. However, when comparing the SDGs indicators and happiness indicators, Iriarte and Musikanski (2019) found that relevant aspects for happiness such as the subjective perspective and indicators related to time balance, community and social support, subjective well-being and health are not included directly within the SDGs indicators. The study indicated that the SDGs framework lack the means to measure happiness and well-being, and more research is needed to understand how the process of moving toward the SDGs will affect happiness and well-being levels.

Various authors have proposed that the SDGs are the best framework to create happier societies (Sachs, 2018). Still, there are many questions about the intersections between the SDGs progress and happiness levels that need to be explored. This paper aims to understand the intersections between the SDG score, well-being level, and ecological footprint (EF) at the country level. For that, this study provides a holistic analysis between the happiness and well-being levels with the SDGs index and various ecological indicators to better understand the interactions among them.

7.2 Methods

To explore the interactions between sustainable development, environmental impact and subjective well-being (SWB) at the country level we carried out a correlation analysis. For that, we have selected the following indicators at country level:

- SWB as reported by Helliwell et al. (2020)
- Sustainable development index (the SDGs score as reported by Sachs et al., 2020)
- Key environmental impact indicators: total EF of production; total EF of consumption; Ecological balance of deficit or reserve (as reported by the Ecological Footprint Network, 2020); and Spillover Score (based on the score provided by Sachs et al., 2020).

The statistical analysis was based on Pearson's Correlation run with SPSS. Before conducting the analysis, the non-linear data were transformed, and normal distributions were established. Next, variables were correlated and within given significance levels.

7.3 Results and Discussion

The results of the correlations between SWB, sustainable development index and environmental impact at the country level is summarized in Table 7.1 and Fig. 7.1.

We found a significant correlation level between the SWB score and the SDGs index, i.e., countries with higher SDGs index score tend to do better in terms of SWB, which reinforceswhat De Neve and Sachs (2020) had found. Despite the overarching positive correlation between the SWB and the SDGs index, De Neve and Sachs (2020) identified tensions between SWB and SDG 12 (sustainable consumption

Table 7.1 Correlation analyses of subjective wellbeing (SWB) and sustainable development goals (SDGs) score on various environmental impact factors

		SDGS index score	Total EF (Production)	Total EF (Consumption)	Ecological balance	Spillover score
SWB	Pearson's correlation	0,78**	0,62**	0,69**	−0,17*	−0,67**
	N	143	144	144	144	143
SDGS index score	Pearson's correlation		0,53**	0,56**	−0,09	−0,52**
	N		163	163	163	163

* represents significance at 10% level. ** represents significance at 5% level

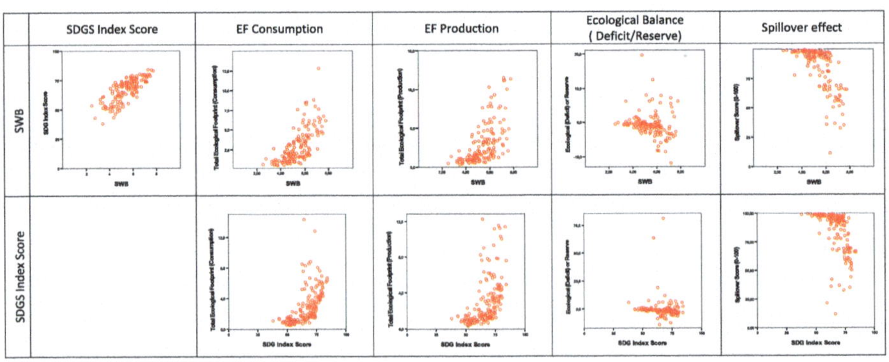

Fig. 7.1 Subjective well-being and SDGS score on various environmental impact factors

and production) and SDGS 13 (climate change) which might occur where actions needed to achieve sustainability may challenge people into changing behaviors and potentially reducing their well-being (at least in the short term). However, their results also suggest that, in the long term, the outcome of environmental policies is positively correlated with SWB which points the relevance of environment on happiness. Also, pro-environmental behavior can enhance SWB (Kasser, 2017).

We also found that the higher the EF of Consumption and the EF of Production, the higher the SWB and the SDGs score. These results are aligned with those of Knight and Rosa (2011) who found a positive correlation between average life satisfaction and EF. Looking at the Ecological Balance (deficit/reserve), results showed that the Ecological Balance was only relevant for the SBW and not for the SDGS index score. Regarding the spillover effects, the higher the SWB and the SDGS score the higher the spillover score by country. Sachs et al. (2020) report that high-income countries generate the largest negative spillovers, which undermine other countries' efforts to achieve the SDGs.

Ecological Footprint and spillover effects are deeply correlated with the levels of consumption. Applying current consumption rates, the universal achievement of high life satisfaction would require a level of resource use that is 2 to 6 times the sustainable level (O'Neill et al., 2018), which violates the planetary boundaries, the safe operating space on earth.

The association between well-being levels and consumption levels is not clear. While some studies observe that a decrease in consumption level is associated with a decline in happiness (Fanning & O'Neill, 2018) other authors (Brown & Kasser, 2005) argue that the ecological well-being (which would imply that we reduce our consumption to reduce environmental impacts) and SWB do not have to be contradictory pursuits and even that lower personal footprints do not reduce well-being (Verhofstadt et al., 2016). These apparently divergent visions might be explained by the wellbeing-consumption paradox which proposes that well-being and consumption levels are positively correlated until a plateau point where increasing consumption does not significantly increase wellbeing (Fanning & O'Neill, 2018). Another explanation may be the importance of how much one consumes relative to others and its perceived value, which varies amongst different people.

The SDGs propose a holistic view for sustainable development with social, economic and environmental considerations—the goals are conceptually intertwined which implies that synergies and trade-offs occur among them. For example, the SDGs aspire to economic growth (see target 8.1) and at the same time they aim for sustainable production and consumption (goal 12) and environmental sustainability (goals 13–15) which are not consistent (Hickel, 2019) given a lack of conclusive evidence that economic growth is decoupled from material consumption (Agnolucci et al., 2017). If the mechanisms to decouple economic growth from environmental impact are not effective by 2030, the tensions generated might impact well-being.

Even if relevant progress is achieved in the SDGs, there are important aspects related to happiness and wellbeing that won't be measured. The SDGS target 17.19 considers the development of measurements of progress of sustainable development that complement gross domestic product (GDP), which might help to introduce happiness and well-being measures to a broader audience. Inspired by this, the idea of an additional SDG 18—*Happy Societies*—was envisioned to specifically include important indicators for happiness, which are not well represented within the SDGs. Based on the gap analysis between the happiness and well-being indices and the SDGs, Iriarte (2019) proposed a SDG 18 articulated in two targets: (i) generate the conditions for all people to be happy and (ii) improve people's life perceptions. The first target captures relevant issues to happiness that are not captured by the SDGs such as working hours or life expectancy while the second target include subjective perceptions about health, relationships, etc.

While the world as a whole has made progress on the SDGs, Covid-19 has severe negative implications for most SDGs (Sachs et al., 2020). *Six transformations to achieve the sustainable development goals* was proposed in 2019 by Sachs et al., then emphasized in 2020, to catalyze the progress towards the SDGs. Proposed transformation number two considers health, well-being and demography. In this proposal, the authors suggested including policies and metrics to raise well-being and quality of life and SWB in policy design.

Beyond the themes and indicators considered in both frameworks, the 2030 Agenda has contributed to transcend the *thematic silo* approach. This holistic view of development paves the way for other holistic processes, such as the happiness and well-being metrics, to be more understandable by a broader public.

The higher the SDGS score, the higher the SWB index and vice versa. To assure that the SDGs work for well-being now and in the future, we need to work to and research how to decouple them from environmental, economic and social negative impacts. With the aim of better integrating the concepts of sustainability and happiness various concepts have been coined. The *Sustainability Through Happiness Framework*, for instance, aims to introduce happiness visioning in sustainability interventions (Cloutier & Pfeiffer, 2015). The concept *sustainable happiness* has been understood as the "happiness that is achieved without the exploitation of other people, the depletion of non-renewable resources, and the wellbeing of future generations" (O'Brien, 2005, p. 18) which might be an inspiring definition for further work. A critical challenge for sustainable happiness to occur is the need to either improve the decoupling of happiness and material consumption, or happiness levels that are less sensitive to declining consumption (Fanning & O'Neill, 2018).

Sustainability transitions will be easier to implement if they also increase individuals' well-being (Verhofstadt et al., 2016) or at least not compromise its levels, since lowering well-being erodes the support for incumbent governments (Ward, 2020). We already know about the poor outcomes of the Rio Principles and the Agenda 21, which provides great opportunity to question the cosmovision, mindsets, values, beliefs, fears and attitudes behind. A way to promote sustainable happiness is to promote more conscious societies by means of encouraging intrinsic values and motivations as well as mindfulness (Brown & Kasser, 2005).

As Harari stated during the pandemic, "Humanity has all the scientific knowledge and technological tools to overcome the virus. The really big problem is our own inner demons, our own hatred, greed and ignorance" (The Dong-A Ilbo, 2020). He stressed the importance of avoiding hatred and embracing compassion and solidarity in combating Covid-19. As every crisis, the Covid-19 pandemic is also an opportunity to learn to respond with compassion, generosity and wisdom, which contribute to a sustainable society.

This study has tried to develop a holistic view of relevant intertwined issues related to SDGs, Ecological Impact Indicators and Happiness and Well-being. A better understanding of the relation between the behaviors aligned to a sustainable ecological footprint and its relation with well-being is key to progress to sustainability. A deeper comprehension of the role of intrinsic motivations and consciousness is key to that.

7.4 Conclusions and Way Forward

The SDGs take into account the social, economic and environmental dimensions of sustainable development so synergies and trade-offs are generated. Therefore, we need to put attention as well as resources and will to avoid tensions in the outcomes that could generate impacts on happiness and well-being. We should reinforce synergies and avoid trade-offs between sustainability levels and happiness and well-being levels. A key point is to understand how to decouple well-being levels from consumption levels and respective environmental impacts.

We have explored the interactions between the SDGS index, subjective well-being, and various environmental indicators at the country level and have observed that the higher the SWB, the higher the SDGs score, the environmental footprint and spillovers effects.

A way for the SDGs to reinforce the promotion of happier societies is if they explicitly complement the battery of indicators with happiness and well-being metrics. This could help with better monitoring and evaluation of the well-being outcomes in the journey toward sustainable development.

There are many examples of international processes that don't get the results they were designed for and offer an extraordinary opportunity to extract lessons. Other sources of inspiration might be some definitions and terms such as sustainable happiness. A popular saying states "If you want different results, you have to try different approaches"—new mindful visions focused on happiness and well-being articulated by intrinsic values and motivations might be the way toward a sustainable future. In this, compassion, generosity and wisdom can pave the way.

References

Agnolucci, P., Flachenecker, F., & Söderberg, M. (2017). The causal impact of economic growth on material use in Europe. *Journal of Environmental Economics and Policy, 6*(4), 415–432.

Brown, K. W., & Kasser, T. (2005). Are psychological and ecological well-being compatible? The role of values, mindfulness, and lifestyle. *Social Indicators Research, 74*(2), 349–368.

Brundtland, G. (1987). *Report of the world commission on environment and development: Our common future*. United Nations General Assembly document A/42/427.
Cloutier, S., & Pfeiffer, D. (2015). Sustainability through happiness: A framework for sustainable development. *Sustainable Development, 23*(5), 317–327. https://doi.org/10.1002/sd.1593
De Neve, J. E., & Sachs, J. D. (2020). Sustainable development and human well-being. In J. F. Helliwell, R. Layard, J. D. Sachs, & J. E. De Neve (Eds.), *World happiness report* (pp. 113–128). Sustainable Development Solutions Network. https://worldhappiness.report/archive/
De Neve, J. E., Diener, E., Tay, L., & Xuereb, C. (2013). The objective benefits of subjective well- being. In J. Helliwell, R. Layard, & J. Sachss (Eds.), *World happiness report* (pp. 54–79). Sustainable Development Solutions Network. https://worldhappiness.report/archive/
Durand, M., & Exton, C. (2019). Adopting a well-being approach in central government: Policy mechanisms and practical tools. In *Global happiness and wellbeing policy report* (pp. 141–161).
Ecological Footprint Network (2020). *Data and methodology*. https://www.footprintnetwork.org/resources/data/
Fanning, A. L., & O'Neill, D. W. (2018). The wellbeing-consumption paradox: Happiness, health, income, and carbon emissions in growing versus non-growing economies. *Journal of Cleaner Production, 212*, 810–821. https://doi.org/10.1016/j.jclepro.2018.11.223
Helliwell, J. H., Huang, H., Wang, S., & Norton, M. (2020). Social environments for world happiness. In J. F. Helliwell, R. Layard, J. D. Sachs, & J. E. De Neve (Eds.), *World happiness report* (pp. 13–46). Sustainable Development Solutions Network. https://worldhappiness.report/archive/
Hickel, J. (2019). The contradiction of the sustainable development goals: Growth versus ecology on a finite planet. *Sustainable Development, 27*(5), 873–884.
IPSOS. (2020, October). Global happiness 2020: What makes people happy in the age of COVID-19. *A 27-country global advisor survey*. https://bit.ly/2In3Hjt
Iriarte, L. (2019). SDG 18: Happy societies. *El Buen Vivir*. https://elbuenvivir.org/estudios-e-investigaciones/
Iriarte, L., & Musikanski, L. (2019). Bridging the gap between the sustainable development goals and happiness metrics. *International Journal of Community Well-Being, 1*(2), 115–135.
Kasser, T. (2017). Living both well and sustainably: A review of the literature, with some reflections on future research, interventions and policy. *Philosophical Transactions of the Royal Society a: Mathematical, Physical and Engineering Sciences, 375*(2095), 20160369.
Keeble, B. R. (1988). The brundtland report: Our common future. *Medicine and War, 4*(1), 17–25.
Knight, K. W., & Rosa, E. A. (2011). The environmental efficiency of well-being: A cross-national analysis. *Social Science Research, 40*(3), 931–949.
Nilsson, M., Griggs, D., & Visbeck, M. (2016, June 15). Policy: Map the interactions between sustainable development goals. *Nature News, 534*(7607), 320–322. https://www.nature.com/news/policy-map-the-interactions-between-sustainable-development-goals-1.20075
O'Brien, C. (2005, June). Planning for sustainable happiness: Harmonizing our internal and external landscapes. *Rethinking development: 2nd international conference on gross national happiness* (pp. 1–22). http://www.gpiatlantic.org/conference/papers/obrien.pdf
O'Neill, D. W., Fanning, A. L., Lamb, W. F., & Steinberger, J. K. (2018). A good life for all within planetary boundaries. *Nature Sustainability, 1*(2), 88–95. https://doi.org/10.1038/s41893-018-0021-4
Organization for Economic Cooperation and Development (OECD). (2018). *Measuring distance to the SDG targets 2017: An assessment of where OECD countries stand*. https://doi.org/10.1787/9789264308183-en
Organization for Economic Cooperation and Development (OECD). (2019). *Accelerating climate action: Refocusing policies through a well-being lens*. https://doi.org/10.1787/2f4c8c9a-en
Sachs, J. D. (2018). Good governance in the 21st Century. *Global happiness policy report 2018* (pp. 3–10). https://www.happinesscouncil.org/report/2018/
Sachs, J. D., Schmidt-Traub, G., Kroll, C., Lafortune, G., Fuller, G., & Woelm, F. (2020). The sustainable development goals and COVID-19. *Sustainable Development Report, 2020*. https://sdgindex.org/reports/sustainable-development-report-2020/

Sachs, J. D., Schmidt-Traub, G., Mazzucato, M., Messner, D., Nakicenovic, N., & Rockström, J. (2019). Six transformations to achieve the sustainable development goals. *Nature Sustainability, 2*, 805–814. https://doi.org/10.1038/s41893-019-0352-9

Stakeholder Forum for a Sustainable Future. (2012). Review of implementation of Agenda 21 and the Rio Principles: Synthesis. *Sustainable Development in the 21st century (SD21)*. https://bit.ly/37LicGn

Stiglitz, J. E., Fitoussi, J. P., Durand, M. (2019). Measuring what counts: A new dashboard for well-being. The New Press.

The Dong-A Ilbo. (2020, April 24). *Harari warns against hatred and ignorance in fighting COVID-19*. https://www.donga.com/en/article/all/20200424/2046040/1/Harari-warns-against-hatred-and-ignorance-in-fighting-COVID-19

United Nations. (2015). Transforming our world: The 2030 agenda for sustainable development, A/RES/70/1. https://sustainabledevelopment.un.org/post2015/transformingourworld/publication

Verhofstadt, E., Van Ootegem, L., Defloor, B., & Bleys, B. (2016). Linking individuals' ecological footprint to their subjective well-being. *Ecological Economics, 127*, 80–89.

Ward, G. (2020). Happiness and voting: Evidence from four decades of elections in Europe. *American Journal of Political Science, 64*(3), 504–518. https://doi.org/10.1111/ajps.12492

Chapter 8
Are Turkish Housewives Happy? A Qualitative Approach

Shoirakhon Nurdinova and Funda Rana Adaçay

Abstract Individual work life satisfaction and individual happiness affect both the employed and non-employed. In terms of economic growth, labor factors associated with the production process increase income and welfare. In this context, a related increase in the quality of life also affects individual life satisfaction. Conversely, some women prefer never to enter the work force, thus making them happier to contribute to their homes. Such outcomes might explain situations like gender discrimination in work life, inequality in wages, or home-work conflict. This study analyzes the happiness level of Turkish housewives, who, according to data of the Turkish Statistical Institute, numbered 11 million in 2018. According to data from *the Social Values Survey*, the happiness level of housewives is relatively high compared to other categories (unemployed, students, retired, self-employed, part-time/full-time employed). A qualitative study using semi-structured interviews with 60 housewives from different regions of Turkey explores factors affecting women's decisions to participate in the labor market, or not. Among 60 respondents, 43 were happy to be a housewife, while 17 wanted to work. Results indicate that housewives believe the primary reason to work would be for income, although it might lower their overall happiness. Findings reveal that there are two gender-based reasons for housewives' staying home: 'child/elderly care;' and 'men do not allow women to work'. These responses affirm socio-cultural norms assigning responsibility for elders and child-care to women. We suggest that future studies should study men's perspectives on how to increase life satisfaction/happiness levels of working women in Turkey.

This study was supported by Anadolu University Scientific Research Projects Commission under the grant no 1709E494. This book chapter is derived from the authors' book entitled "Türkiye'de Kadın İstihdamının Mutluluk Ekonomisi Açısından Değerlendirilmesi".

S. Nurdinova (✉)
Namangan State University, Namangan, Uzbekistan
e-mail: sh.nurdinova@gmail.com

F. R. Adaçay
Anadolu University, Eskişehir, Turkey
e-mail: frozbey@anadolu.edu.tr

S. Cloutier et al. (eds.), *Linking Sustainability and Happiness*,
Community Quality-of-Life and Well-Being,
https://doi.org/10.1007/978-3-030-89559-4_8

Keywords Happiness · Housewife · Turkey · Qualitative study

8.1 Introduction

According to data from the Turkish Statistical Institute, Turkey has a low rate of female labor participation. The size of the non-institutional civilian population of Turkey is approximately sixty million, thirty-two millions of whom are active/formally registered in the labor force (TSI, 2018). Data demonstrates that only 32% of working-age women in Turkey are employed, compared to 44.6% in OECD countries or 46% in EU countries (OECD, 2019). Some authors indicate that female labor participation in Turkey is low because social beliefs, norms, and behaviors also play an important role in women's decisions to seek employment (Göksel, 2013; İlkkaracan, 2012).

Figure 8.1 shows that the Turkish female labor participation rate was at its peak level in 1990. In 1993, the rate rapidly declined and rose remarkably in 1994. Since 1990, the female labor participation rate is a fluctuating trend (Fig. 8.1).

The reasons that Turkish women do not participate in the labor force are shown in Table 8.1. It reveals that being a housewife is the most frequent reason for women not

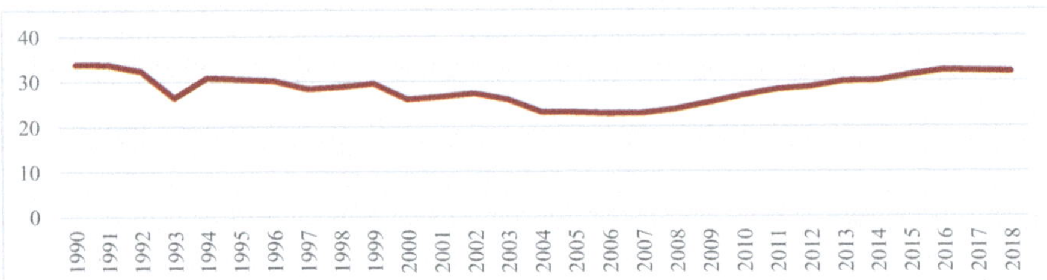

Source: Work Bank data, 1990-2018

Fig. 8.1 Female labor participation rate in Turkey, in %. *Source* Work Bank data, (1990–2018)

Table 8.1 Reasons that women are not participating in the labor force (%)

	2002	2003	2004	2005	2006	2007	2008
Student	15.6	16.9	17.1	17	17.7	16.2	9.4
Housewife	63.4	61.1	64.6	62.6	64.4	65.1	68.5
Retired	3.1	2.5	3.5	3.6	3.5	4.2	6.1
Elderly (60+)	6.9	11.3	6.7	6.9	5.2	5.1	6.4
Disabled	1.3	1.4	1.8	1.9	1.9	2	2.3
Personal Reasons	7.4	4.8	3.9	4.9	4.4	4.3	3.9
Other	2.3	2	2.4	3.1	2.9	3.1	3.4

Source TURKSTAT Household Budget Surveys (2002–2008)

participating in the labor force. According to the Turkish Statistical Institute, there were eleven million housewives in Turkey in 2018. In macroeconomic terms, this number shows that high potential labor is out production, which factors of production are used inefficiently.

According to international indicators, the female labor participation rate is accepted as a measure of gender equality across countries. This rate implies the economic and social empowerment of women, which may account for women's being economically independent. In this paper, we evaluate women's lack of labor participation differently. An analysis of the economics of happiness asks different questions than an analysis focused on labor productivity, and we consider the happiness of housewives as one criterion in the economics of happiness. Happiness refers to satisfaction with life-as-a-whole (Veenhoven, 2013).

As seen in Fig. 8.2, the average life satisfaction of housewives is relatively higher than other social categories (full-time, part-time, self-employed, retired, student, unemployed, and others). However, a different profile exists in 2007 and 2012: overall satisfaction has increased, and working people's life satisfaction level has improved more than housewives'. In this period, while the average life satisfaction of housewives and retirees is at the same level, the life satisfaction of full-time employees is highest. According to 2012 data, there is a decrease in the average life satisfaction of housewives—housewives also have slightly lower life satisfaction than full-time, part-time, and self-employed categories.

This paper focuses on the happiness of Turkish housewives and examines the factors affecting women's decisions to not participate in the labor force. Instead of conducting experimental studies or large-scale surveys, qualitative research techniques were used to reveal the perceptions of housewives about working and their happiness. The questions asked are related to the happiness of the housewives, work and family, sources of happiness, and the relationship between work and happiness.

While there are numerous studies on happiness and the employment status of women, few studies have investigated a deeper understanding of the happiness of

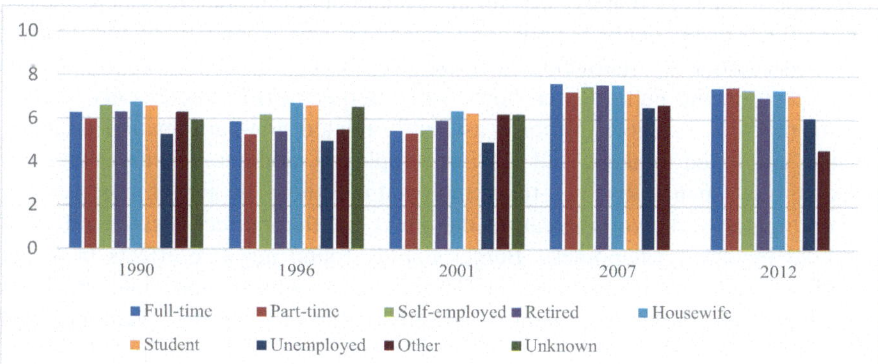

Fig. 8.2 Life satisfaction according to employment status. *Source* World Values Survey (1990–2012)

housewives using a qualitative approach. To our knowledge, no studies have explored the happiness of Turkish housewives, although some have used *housewife* as a dummy variable in an econometric model.

8.2 Literature Review

Although the concept of happiness can be traced back to the Ancient Greeks, empirical studies of happiness began in 1974 with Richard Easterlin's seminal work on happiness and economic growth, and these studies rapidly increased in the 1980s (Easterlin, 1974). In this paper, we refer to the categories of women who are full-time working women, part-time working women, unemployed women, and full-time housewives. We defined a housewife as a woman doing unpaid household chores for others, such as caring for the sick or elderly, regardless of marital status. We excluded women who work seasonally or create crafts at home for sale. Full-time working women refer to the employed who work more than thirty hours per week, while part-time working women are employed in jobs which are less than thirty hours per week. We defined unemployed women as those who do not have a job, but who actively looked for work in the past four weeks, and are currently available for work.

We divided empirical studies on the economics of happiness showing the relationship between the level of happiness and employment status according to the happiness of housewives:

1. Housewives are happier than other social categories
2. Housewives are less happy than other social categories
3. Housewife is happier/less happy than some social categories.

Hall and Gordon's (1973) study shows that full-time workers are the happiest and the most satisfied as compared to housewives and part-timeworking women. Mencarini and Sironi's (2012) demonstrated similar findings and reveal that being engaged in housework that exceeds the median amount time recorded in a specific country affects respondents' happiness in a negative way: women living in a country where partners usually share equally household chores feel more unhappy if they have to perform the bulk of total housework. Berger (2013) studies only women who are mothers, categorizing them as family-related non-participation mothers, part-time employed mothers, and full-time employed mothers. The first two categories are less satisfied than full-time employed mothers. The author indicates that not working/part-time working causes the direct and the indirect negative effect of household income on happiness. Berger's study found negative effects from the institutional and social conditions in Germany that prevent many mothers from reconciling (full-time) employment with motherhood. Another study demonstrates that the life satisfaction of working women is relatively higher compared to non-working women in India (Totuka, 2013). Gokdemir and Tahsin's (2014) research on factors affecting the life satisfaction of Northern Cypriot women demonstrates that housewifery negatively affects the life satisfaction of women.

Ferree (1976) argues that full-time working women are happier and feel better than full-time housewives. Part-time workers are satisfied with their situation and most interested in the non-financial aspects of their jobs. Disagreeing Ferree's study, Wright (1978) disclaims that working women are happier and more satisfied than full-time housewives. His study, based on data from the University of Michigan and the National Opinion Research Center between 1971 and 1976, shows that both working outside the home and full-time housewifery have benefits and costs attached to them; the net result is that there are no consistent or significant differences in patterns of life satisfaction between the two groups. Okulicz-Kozaryn and da Rocha Valente's (2018) study using data from US General Social Surveys from 1972 to 2014 supports Wright's study. The results indicate that housewives and part-time working women are slightly happier than full-time working women. The model used in the study estimates control variables such as age, marital status, class, having children, etc. Furthermore, Michoń's (2014) study on the relationship between women's employment status and life satisfaction includes Hungarian women aged 20–59 based on data from the World Value Survey. The study concludes that housewives are happier than full-time workers and unemployed mothers.

Graham et al.'s (2004) paper on income, health, and other factors affecting happiness, using panel data from Russia, partly supports previous studies. Results show that housewives are relatively less happy than the self-employed and students, while they are happier than retired and employed people. Arshad et al.'s (2015) study compares life satisfaction in working women and non-working women of Pakistan. Findings reveal that working women are more highly satisfied than non-working women in Faisalabad, while non-working in Islamabad women are more satisfied. Some studies show that the economic performance of countries affects the happiness level of working women and housewives. For instance, an empirical study on the happiness of working women and housewives was estimated by a regression model in different groups such as upper-income economies, middle-income economies, and low-income economies. Findings reveal that there is no difference in the happiness level between housewives and working women in the upper- and low-income economies, while part-time working women are relatively happier than housewives and that both part-time working women and housewives are happier than full-time working women in middle-income economies (Beja, 2014). Furthermore, participation in the labor force out of economic necessity might negatively affect happiness. For instance, Orden and Bradburn's (1969) study revealed that working women's happiness level decreases when they are in the labor market only out of economic necessity, while part-time working women at the labor market out of economic necessity are more satisfied than full-time working women. Also, the study demonstrates part-time working women are relatively happier than full-time working women and housewives.

While reviewing the literature on the happiness of women, the results of empirical studies reveal many differences. These differences may be caused by the cultural structure or methodology, and/or country or region included in the study. Our study is a contribution to the field of happiness studies as qualitative research in Turkish housewives. In this study, distinguishing factors stimulating Turkish women to choose to be a housewife include culture and socio-demographic factors, but these alone do not account for the decisions that Turkish women make to engage or not engage in the labor market.

8.3 Methodology

To investigate a deeper understanding of the views of the housewives about employment and happiness, an in-depth interview was conducted. The author conducted 60 semi-structured in-depth interviews; informants provided oral consent. Hagaman and Wutich (2017) suggest that 20 to 40 interviews are enough to reach data saturation for finding cross-cultural meta themes. The study includes six regions, districts, and villages: Sanliurfa, Konya, Istanbul, Izmir, Trabzon, and Eskisehir. The researcher, Dr. Nurdinova conducted face to face interviews in Turkish; interviews were translated into English word by word. Respondents were selected based on different class/ethnicity statues and included women with and without children. The researcher visited the houses of the respondents and met in public places such as parks and cafes, and created a research atmosphere for interviewees to feel comfortable and not under pressure from family members. The interview consisted of open-ended and closed-ended questions about the socio-economic characteristics and happiness of the women. The main focus of the interviews was the importance of working, happiness level, sources of happiness, satisfaction of household income, reasons for not working, and perceptions about working women-motherhood. It should be noted that the interview form was designed to maintain consistency in interviews with different respondents and to help them focus on important issues in the interview process.

After interviewing two respondents in a pilot study, minor edits were made to the questions. To obtain more detailed answers, the decision was made to always conduct interviews in an environment where respondents feel more comfortable and are ensured the anonymity. Respondents were selected by snowball sampling. Additionally, the researcher ensured respondents represented the profile of a housewife in the region. Respondents voluntarily participated in the research and all interviews included women aged 15 and over. Some respondents felt happy to express themselves. For instance, when a respondent from Istanbul was interviewed and audio-recorded, she remarked that she felt excited and happy as she gave an interview to Turkish Radio and Television. This example shows that housewives who are silent, but represent a large part of the society, are delighted to express themselves.

The results of qualitative research do not have the feature of *generalizability* as might be more common in quantitative research. The results also do not directly represent the situation, the event, or the overall population (Yıldırım, 1999). A narrow conception of a theory-driven, hypothesis testing, or generalization does an injustice to the variety of contributions that qualitative research can make (Peshkin, 1993). Rather than drawing on validity and reliability tests commonly used in quantitative studies, criteria such as credibility, transferability, and validation are taken into consideration in qualitative research (Arastaman et al., 2018). In this study, the researcher used methods such as long-term interaction, depth-focused data collection, variation, and expert examination to have the features mentioned above.

Respondents were de-identified to conduct the interview anonymously and coded to relate to their region like Trabzon1, Izmir3, etc. Interviews lasted an average of 35 min and were audio-recorded. Additionally, interview notes from an interview visit are used in the study. Computer-aided qualitative data analysis software—NVivo8 was used to analyze, encode, and generate data.

8.4 Results & Discussion

In terms of marital status, 48 interviewees were married, 3 were single, 5 were widowed, 3 were divorced and 1 interviewee was the second wife, living without official marriage. 53 interviewees had children. Respondents were between 15 and 65 years old. Table 8.2 shows that the majority of respondents were between 41 and 54, the average age of respondents was 43.5 years old.

Educational level is more likely to promote/correlate with women's participation in the labor force. This finding supports studies indicating a positive relationship between an increased level of education and women's participation in the labor force (Cameron et al., 2001; Standing, 1978). Since women's participation in the labor

Table 8.2 The sample for qualitative analyses

Variable	Categories	Number of respondents (%)
Age	15–28	7 (11.66)
	29–40	16 (26.67)
	41–54	25 (41.67)
	54–65	12 (20.00)
Having children	Yes	53 (88.33)
	No	7 (11.67)
Marital status	Single	3 (5.00)
	Married	48 (80.00)
	Divorced	3 (5.00)
	Widowed	5 (8.33)
	Living without official marriage	1 (1.67)
Educational level	Tertiary education completed	9 (15.00)
	High school	18 (30.00)
	Secondary	9 (15.00)
	Primary	21 (35.00)
	No education	3 (5.00)

force is directly related to the level of education, interviewees were asked about their educational level. In terms of education, 9 interviewees had higher education or an associates degree, 18 had attended high school or professional school, 9 had completed secondary school, 21 had finished primary school and 3 were illiterate.

Working is important in terms of the economic and social empowerment of women. In Turkey, a large proportion of women in the labor market participate in informal labor such as unpaid family workers and working in the agricultural sector. In this context, respondents were asked whether they have worked in the last five years. Most respondents stated that they did not work or worked before and later became a housewife. Forty-six respondents reported that they did not work at all, while twelve respondents reported that they worked a short time before marriage and that they had subsequently left work. Two participants stated that they worked out of economic necessity after marriage, later quitting their jobs. Below, we share the results of the interviews in the following categories: reasons for not working, social pressures, education levels and working, the importance of working, socialization, and happiness sources.

8.4.1 Reasons for Not Working

Respondents were asked why they do not work, and seventeen respondents share their husbands do not permit them to work. Said a respondent: *My husband won't let me work. He says that his financial income is enough. I should keep the house clean, cook food, serve for him, he does not expect anything from me. He is very opposed to working. (Konya2)*. A 24 year old married woman, who left her higher education after marriage, said: *My husband is very jealous, he did not even allow me to go grocery store in the past, and he does not allow me to work (Trabzon9).* Because of a work-family mismatch after marriage, some respondents preferred not to work and focused on mother/wife roles: *Why should anyone else see my child's first step? I'm a perfectionist mother, if you were working, you would disrupt kids: You would think about that kind of thing (Eskisehir9*). A 52 years old woman, who is not satisfied with her status, explained: *My husband didn't allow me to work, but there is a culture difference between me and other housewives. My husband is a teacher. I see that other housewives have restrictions. I'm very comfortable: I take my credit card whenever I want and I shop, there are no restrictions (İzmir2).*

We also asked respondents whether they would like to work if their household income were sufficient. Forty-four out of sixty respondents would not work, while sixteen of them would like work regardless of household income: *I would like work to get away from home problems, to clear my head (Konya4).* Participants who want to work believe that work is about more than earning money. Their perception of working is self-esteem, social environment, and economic freedom. While respondents are asked why they did not want to work, those who were happy in their housewife roles explained reasons such as professional insufficiency, desire not to depend on any institution, having enough money, social prejudice, and the superiority of domestic roles.

8.4.2 Social Pressures

We asked whether respondents had experienced social pressure because they are housewives. Seven respondents had experienced social pressure for working, while fifty-three of the participants said there was no pressure at all: *Nobody interferes, my family and my husband's family are not very supportive about working, they say that men should work and they not say women have to work. My husband's sisters also do not want to (Şanliurfa7).* Living in rural or uran areas also affects women's working: *There is no pressure. As I lived in the countryside, nobody interferes with my life (Konya2).*

Some respondents felt pressure from family members, spouses, friends, and others. The interesting point found that as the level of education of the respondents increases, the possibility of social pressure for working would increase. A respondent said: *There is pressure from my husband, he says that if I worked, we would earn more money (Konya5).* Two respondents stated that they had to work after marriage, while they preferred to be housewives: *I worked at the dorm for a month, because of the financial situation. I started to work because my husband lost his job. My kids were miserable when I worked. I said to my mom, look, one stayed in school, someone stayed there. No one came and cared, no one helped. Then my husband found a job, and I immediately left my job (Sanliurfa7).*

8.4.3 Education Levels and Working

Educated women did not feel less pressure from their husbands to stay at home. Interviewees with a low level of education stated that they did not feel pressure to work and their husbands not permitting them to work. Some interviewees without higher education prefer to be housewives rather than working. Interviewees without a university degree are more likely to be less satisfied with salaries offered for low-level jobs. From a cost–benefit perspective, it is less satisfactory working at unskilled jobs for minimum wage: *If I had graduated a college, I could have worked. I could have hired a baby-sitter. If I had worked as a teacher, I'd been with my kid afternoon. But what's the point if I work for the minimum wage and give it to a babysitter now? How useful can it be to my child? I'm sitting at home and taking care of my kid myself. But now I don't want to work at unskilled works (Izmır4).* The interesting finding is that participants who had high education levels negatively evaluate working women on motherhood. These perceptions might be caused by these participants' experiences due to family-work imbalance.

8.4.4 The Importance of Working

Respondents were asked to explain the importance of working. Analyzing data brings out seven different sub-categories. A majority of respondents mentioned money, income, and the materiality of working. These findings demonstrate that material benefit is one of the specified characteristics and importance of working. An interviewee said: *Working is important only in economic terms, material benefit. I mean, otherwise, why should she work? Why should a woman work if her husband handled it? (Sanliurfa5).* Interviewees who previously worked underlined the satisfaction of working in terms of economic freedom: *You do not have to ask money from your husband for your needs. Working has such advantages (Konya4).* Participants' perceptions of working reveal positive feelings like self-esteem and self-confidence: *Working is very important in terms of economic independence and self-esteem. You contribute to your family. (İzmir9).* Some respondents believe that working brings self-confidence to women: *You feel self-confidence and meet your financial needs (Trabzon3).* Furthermore, freedom is seen as one of the most essential perceptions on the importance of working: *Working is good in terms of economic freedom. I mean, it is different taking money from your husband to buy something than saying that I bought it (Eskisehir4)* (Fig. 8.3).

Studies show that there is a positive and causal relationship between economic freedom and happiness (Gropper et al., 2011; Jackson, 2017). When respondents were asked whom they were economically dependent on—the most frequent responses related to their husbands. 48 respondents were dependent on their husbands, 3 respondents were dependent on their families, 3 respondents were dependent on their father, a respondent was dependent on a daughter, while 8 respondents were economically independent. Respondents were also asked whether they had additional income: in response, 27% indicated that they received additional income from renting apartments, lands, and heirship, while 73% had no additional income.

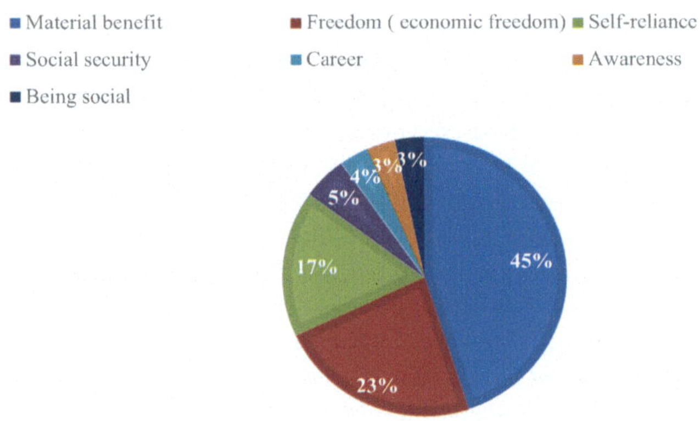

Fig. 8.3 The importance of working (%)

8.4.5 Socialization

Socialization is another frequently mentioned perception of women to explain the importance of working: *It is important in terms of socialization, it is important to be a useful mother to her children. If your spouse had died, then it is important to show yourself to your children as an independent mother (Konya2).* Working is precepted as getting more income from participants with low educational levels, while higher educated participants evaluate working as self-esteem and self-confidence.

8.4.6 Happiness Sources

The interview questions were designed to explore the most important happiness sources of respondents (see Table 8.3). The results showed that children were happiness sources of 32 respondents out of 53 that had children. A spouse was the direct happiness source for 32 respondents; grandchildren—five respondents; herself—three respondents; relatives—three respondents; and friends—one respondent.

As is seen in Table 8.3, the family and success of children are the main happiness sources of housewives. While family and children are the most important factor of housewives, we asked housewives whether they think of working women's attitudes on motherhood and family roles. Fifty-seven percent of the respondents believe that working women could be a more conscious mother who spends productive time with their children: *Working women do not have as much time as a housewife, but I think that they spend more productive time with their children (Konya1).* Participants believe that working is more advantageous for mothers in raising children in the rapidly developing technology age.

Respondents were asked whether they were happy being a housewife. In response, 43 out of 60 respondents said that they are happy being housewives, while 17 of them are not satisfied with their status. According to a respondent, spending every day doing the same household chores is boring: *Actually, I'm not happy being a*

Table 8.3 Happiness sources

Who?	What?
Children (32)	Family (12)
Spouse (16)	The success of children (11)
Grandchildren (5)	Health (10)
Herself (3)	Peace (10)
Relatives (3)	Material values, gold… (8)
Friends (1)	Beliefs, religion, etc. (3)
	Travelling (3)
	The behavior of people (3)

housewife. I think I'd be happier if I worked. I do the same things every day at home, too boring (Konya1). Another respondent compares her life and working women's life: *We see that working women do household chores once a week, we do every day, we cook at home every day. Because they work, their husbands help them. Our life is different- every day we get up in the morning and fulfill chores until evening at the house: cooking, laundry (İzmir2).*

8.4.7 Work-Life Balance

Forty-three percent of respondents think working means that a mother could not express the same affection and would neglect her children. They believe that children of working women are lacking the love of mothers: *Working women cannot be a good mother, her children always remain at halfway. In other words, the working mother's child always expects love from others, it is a big difference between growing up with others and growing up with a mother's love. Working women cannot see the best days of their children (Izmir4).* It is seen that married women decide not to work after having children, given the problems within marriage increasing with children. When the effect of motherhood ties to children is considered, it can be concluded that married women with children are more reluctant to participate in business life. Some participants believe that housewives are better mothers and they are validating their role, which adds to their feeling of happiness: *I am happy, why? I raised my children myself (İzmir3).* Trabzon7 said that she is happy being a housewife: *If I had worked, I could not do household chores: my children are picky.*

Respondents said: *Working life is very busy: it also makes you happy, but the beautiful thing about housewifery is that you have time for yourself: you can participate in social activities, you might be engaged in cultural activities. You have the opportunity to talk to your relatives and your friends more. More opportunities to see around, traveling. When you work, you stay very committed to the business environment, so you are deprived of many things. To avoid that deprivation, I prefer to leave my work (Istanbul2).*

Some respondents interpret working as having negative impacts on marriage. They believe that even if a working woman appears as educated and economically independent on one hand, on the other hand, it might increase the rate of divorce. A working woman has multiple roles in work and at home and may not be able to fulfill her family roles due to fatigue, stress, and a lack of time. They affirm traditional gender roles: *Our Turkish people always wait for women to do domestic chores, so it is difficult for women (Konya3).* Participants believe that women are expected to do household chores, caring for children and elders even though they are in the labor market: *My daughter is a teacher, she works part-time. If a woman full-time employee got tired, would her husband help her? It is awaited from women, not only in Trabzon but everywhere (Trabzon5).*

Respondents' perceptions on working are likely to relate to income. Forty-four out of sixty respondents would not work if household income were enough, sixteen respondents would like work regardless of household income. As working is correlated with educational level, housewives, who had no higher education, are less satisfied with salaries offered for low-skilled jobs. University graduated respondents had social pressure to work and usually this pressure came from family members. In this study, we tried to show the how happiness of housewives connected to work-life. Respondents mentioned their happiness sources as family and children rather than work-life or career.

8.5 Conclusions

This paper investigates the happiness of Turkish housewives. We used a qualitative approach to amplify the voices of housewives on perceptions about happiness and work life. Semi-structured interviews were conducted with 60 women in Turkish, and then transcripts were translated to English. We used Nvivo software to analyse our data. Results indicate that housewives believe the primary reason to work would be for income. Seventeen respondents are out of the labor market as their husbands do not allow them to work. In general terms, they explain that their child care and domestic chores keep them from working, and in their lives and words, they affirm traditional gender roles. The result may originate from the socio-cultural structure or the patriarchal-cultural structure of male dominance, in which women's primary responsibility is the care of children or the elderly in the family. Some respondents shared the perception of working as having negative impacts on marriage and children. As Turkish housewives' main source of happiness is family (children), the government could give longer paid leaves for parents to have the highest rates of happiness and support a higher labor participation rate.

The findings of this study demand the need for future research. Specifically, the happiness of working and non-working women should be investigated in more detail. To increase the female participation rate in the labor market, future studies should also include men's studies or the relationship between patriarchy and happiness of (non)working women. While the study focused on the happiness of housewives, it points to a problem with the unhappiness criteria of working women. Public or private, economic, social, and cultural policies need to be realized to solve this problem, which exceeds the scope of the study. However, the findings point to the necessity of further studies to be carried out both on a micro and macro scale and to increase the effects of existing ones.

References

Arastaman, G., Fidan, İ. Ö., & Fidan, T. (2018). Nitel Araştırmada Geçerlik ve Güvenirlik: Kuramsal Bir İnceleme. *Yüzüncü Yıl Üniversitesi Eğitim Fakültesi Dergisi, 15*(1), 37–75. https://dergipark.org.tr/tr/pub/yyuefd/491262

Arshad, M., Gull, S., & Mahmood, K. (2015). Life satisfaction among working and non-working women. *European Journal of Research in Social Sciences, 3*(1), 7.
Beja, E. L. (2014). Who is happier: Housewife or working wife? *Applied Research in Quality of Life, 9*(2), 157–177.
Berger, E. M. (2013). Happy working mothers? Investigating the effect of maternal employment on life satisfaction. *Economica, 80*(317), 23–43.
Cameron, L. A., Malcolm Dowling, J., & Worswick, C. (2001). Education and labor market participation of women in Asia: Evidence from five countries. *Economic Development and Cultural Change, 49*(3), 459–477. https://doi.org/10.1086/452511
Easterlin, R. A. (1974). Does economic growth improve the human lot? Some empirical evidence. In P. A. David & M. W. Reder (Eds.), *Nations and households in economic growth: Essays in honor of Moses Abramowitz* (pp. 89–125). Academic Press. https://doi.org/10.1016/B978-0-12-205050-3.50008-7
Ferree, M. M. (1976). Working-class jobs: Housework and paid work as sources of satisfaction. *Social Problems, 23*(4), 431–441. https://doi.org/10.2307/799853
Gokdemir, O., & Tahsin, E. (2014). Factors that influence the life satisfaction of women living in the Northern Cyprus. *Social Indicators Research, 115*(3), 1071–1085. https://doi.org/10.1007/s11205-013-0265-3
Göksel, İ. (2013). Female labor force participation in Turkey: The role of conservatism. *Women's Studies International Forum, 41*, 45–54. https://doi.org/10.1016/j.wsif.2013.04.006
Graham, C., Eggers, A., & Sukhtankar, S. (2004). Does happiness pay? In W. D. Glatzer, S. Von Below, & M. Stoffregen (Eds.), *Challenges for quality of life in the contemporary world: Advances in quality-of-life studies, theory and research* (pp. 179–204). https://doi.org/10.1007/978-1-4020-2903-5_13
Gropper, D., Lawson, R. A., & Thorne, J. T. (2011). Economic freedom and happiness. *Cato Journal, 31*(2), 237–255.
Hagaman, A. K., & Wutich, A. (2017). How many interviews are enough to identify metathemes in multisited and cross-cultural research? Another perspective on guest, bunce, and Johnson's (2006) landmark study. *Field Methods, 29*(1), 23–41.
Hall, D. T., & Gordon, F. E. (1973). Career choices of married women: Effects on conflict, role behavior, and satisfaction. *Journal of Applied Psychology, 58*(1), 42–48. https://doi.org/10.1037/h0035404
İlkkaracan, İ. (2012). Why so few women in the labor market in Turkey? *Feminist Economics, 18*(1), 1–37. https://doi.org/10.1080/13545701.2011.649358
Jackson, J. (2017). Free to be happy: Economic freedom and happiness in US States. *Journal of Happiness Studies, 18*(4), 1207–1229. https://doi.org/10.1007/s10902-016-9770-9
Mencarini, L., & Sironi, M. (2012). Happiness, housework and gender inequality in Europe. *European Sociological Review, 28*(2), 203–219. https://doi.org/10.1093/esr/jcq059
Michoń, P. (2014). Mothers and quality of life in 26 EU countries. In E. Eckermann (Ed.), *Gender, lifespan and quality of life: An international perspective* (pp. 83–98). https://doi.org/10.1007/978-94-007-7829-0_6
OECD. (2019). OECD labour force statistics 2019. OECD Publishing. https://doi.org/10.1787/g2g9fb3e-en
Okulicz-Kozaryn, A., & da Rocha Valente, R. (2018). Life satisfaction of career women and housewives. *Applied Research in Quality of Life, 13*(3), 603–632. https://doi.org/10.1007/s11482-017-9547-2
Orden, S. R., & Bradburn, N. M. (1969). Working wives and marriage happiness. *American Journal of Sociology, 74*(4), 392–407. https://doi.org/10.1086/224664
Peshkin, A. (1993). The goodness of qualitative research. *Educational Researcher, 22*(2), 23–29.
Standing, G. (1978). Labor force participation and development.
Turkish Statistical Institute (2018).
Totuka, N. (2013). Life satisfaction in working and non-working women. *Indian Journal of Positive Psychology, 4*(1), 185.

TURKSTAT Household Budget Surveys, (2002–2008).
Veenhoven, R. (2013). Conditions of happiness. Springer Science & Business Media.
Work Bank data (1990–2018).
World Values Survey (1990–2012).
Wright, J. D. (1978, May). Are working women really more satisfied? Evidence from several national surveys. *Journal of Marriage and the Family,* 301–313.
Yıldırım, A. (1999). Nitel araştırma yöntemlerinin temel özellikleri ve eğitim araştırmalarındaki yeri ve önemi. *Eğitim ve Bilim, 23*(112).

Chapter 9
Inner Happiness and Environmental Preservation in a Green Space Community

Sauwalak Kittiprapas

Abstract The following chapter fills an identified research gap within the existing, but limited, happiness and sustainable development literature. Specifically, the study investigates the relationship between happiness and environmental preservation, where inner dimensions of happiness have not often been considered. The research was conducted within a green-space community in Thailand, Bangkajao, using both qualitative and quantitative approaches and the *Buddhist Sustainable Development* (BSD) framework. Empirical results confirm that people are significantly happier from constructs of inner happiness (i.e., feeling they have enough, giving/helping, lower material desires, and management of suffering), environmental quality, good family and community relationships, time, health, and perception in relative income. Economic variables, such as monetary income and employment, do not significantly affect happiness. Overall, several inner happiness indicators positively affect environmental quality and the happiness derived from sustainably preserving the green-space. Consistently, conserving the quality of the environment and leading moderate life styles are associated with the happiness of community residents. Specifically, inner happiness significantly results in enhanced environmental quality. Results from this study support the *Buddhist Sustainable Development* framework positing that a focus on enhancing inner happiness can promote environmental preservation and sustainable development, and vice versa.

Keywords Happiness · Environment · Sustainable development · Inner happiness · Buddhist sustainable development · Bangkajao

---

S. Kittiprapas (✉)
International Research Associate for Happy Societies (IRAH) and Faculty of Economics, Rangsit University, Pathumthani, Thailand
e-mail: sauwalak.k@rsu.ac.th

S. Cloutier et al. (eds.), *Linking Sustainability and Happiness*,
Community Quality-of-Life and Well-Being,
https://doi.org/10.1007/978-3-030-89559-4_9

9.1 Introduction

Following the *Buddhist Sustainable Development* (BSD) framework (Kittiprapas, 2018a), which highlights inner happiness as a key factor to shift towards sustainability, this article aims to empirically test the relationship between inner happiness and sustainable development. Although there are an increasing number of studies exploring the relationship between happiness and the environment (Krekel & MacKerron, 2020; Welsch, 2002), it is rare to find such studies that empirically test the link between inner happiness (the focus of BSD) and environmental preservation. Thus, linking inner happiness and environmental preservation as a pathway toward a sustainable future is the main contribution of this chapter.

To date, two empirical case studies in Thailand have been conducted to enhance the BSD framework and confirm the importance of inner happiness to overall happiness. (Kittiprapas, 2018b; Kittiprapas, 2020). Results in both cases tend to be consistent; for example, there was little to no effect of income on happiness but more important effects from health and other non-income variables. In addition, the studies demonstrated the crucial role of inner happiness as significant to overall happiness.

The results of the first case study indicate the positive effect giving/helping others has on happiness and the second case confirms the significance of both feeling one has *enough* in material wealth and environmental quality on happiness of all community groups regardless of income levels, age, status, or occupation. This chapter builds on these works by using both qualitative and quantitative approaches to analyze happiness characteristics of the green-space community *Bangkajao*. The intention of this chapter is to provide additional information on the significance of inner happiness to overall happiness and the results of the relationship between inner happiness and environmental preservation. The chapter also tests the relationship of specific happiness outcomes from preserving the natural environment, as Bangkajao offers unique characteristics from preserving a large green-space over many decades.

Bangkajao is important for its environmental impacts, listed as *The Best Urban Oasis of Asia* by *Time Magazine Asia* in 2006. The *Thailand Greenhouse Gas Management Organization* (Public Organization) estimated the Bangkajao green-space produces 6 million tons of oxygen per day and traps 6000 tons of carbon dioxide per year (http://www.suansri-bangkachao.com/en/home/lesson4/get-to-know-bangkachao). Moreover, as the wind flows from the Gulf of Thailand and passes through Bangkajao to Bangkok about 9 months a year and, due to its close proximity to Bangkok, the community is often referred to as the *green lung* of Bangkok (Bouchet, 2013) or the lung of Bangkok (Marshall, 2006).

Finally, it is interesting that the community members have kept and cared for this green-space and the related agricultural sector for such a long time, even though it is surrounded by urban buildings and development within Bangkok and the Samutprakarn province. Consequently, learning about the community members' happiness and ways of life in this green community offers important lessons for the future. Ultimately, this chapter seeks to explore the relationship between Bangkajao community members' happiness and environmental preservation, which could be useful for

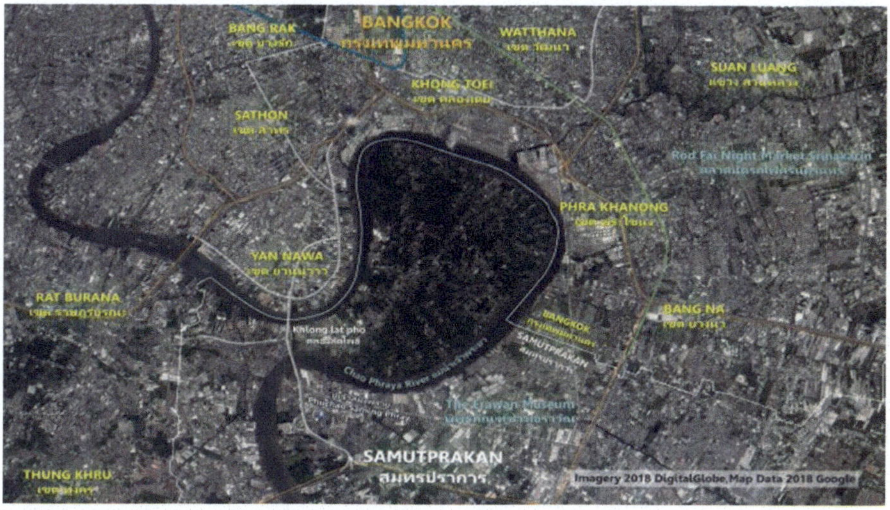

Fig. 9.1 The location and landscape of Bangkajao green-space community. Google map, 2018

guiding pathways toward sustainable development. The following sections provide descriptive information on the community, followed by methods of analysis, a results and discussion of the study, followed by some conclusions.

9.2 Background and Historical Aspects of Bangkajao Community

Bangkajao is a more than century old Buddhist community settlement in the Prapradaeng district of the Samutprakarn province in Thailand. The community has a unique landscape surrounded by water and looks like a green island in the middle of concrete buildings of Bangkok and Sumutprakarn (Fig. 9.1).

Surrounded by the Chao-Praya river, Thailand's main river flowing from the Northern part of the country to the sea into the Gulf of Thailand, the Bangkajao community is located in the opposite side of the river of Bangkok and Sumutprakarn cities. The *Lad Pho* canal is situated at the western end of the green-space, essentially making the community an island. Its total area covers six sub-districts with about 1891 ha (~4672 acres or 11.818 Rai). There are boats that connect the island to several of Bangkok's piers and a road connecting to Bangkok via the Phrapradaeng district of Samutprakarn.[1] Bangkajao has three ecosystems of freshwater, salt water and brackish water. Four to five decades ago, most of the area (about 90%) was devoted to agriculture (growing fruit trees). Over the past two decades, however, the land has

[1] More information about Bangkajao can be found from http://www.suansri-bangkachao.com/en/home/lesson4/get-to-know-bang-kachao.

been developed into more commercial and residential areas, tourism related activities, with increased construction and manufacturing areas from outsiders. Currently, green-space within the entire area has dropped to about 45–50% of the island with a remaining variety of fruit trees. Nowadays, the green island is a popular place for eco-tourism and bicycling, with a large public park and natural learning center, many fruit tree gardens and canals, home-made products and food, local markets, and many traditional temples.

Historically, there were plans to expropriate the whole island to the government and relocate local people to reside in another province or in flat buildings. Local people did not agree and fought to live on the land, where they had preserved the environment by themselves for a long time and intend to live in a similar way. Ultimately, they made a formal petition to the Royal office, which resulted in a negotiation and an agreement that the government can buy some pieces of land from those who were willing to sell and those who prefer to live on their homeland would stay. Thus, the community is a mix of private owned land and some public land (bought from those willing to sell in the past). Some public lands have become a large public park, while a lot of small pieces of public land are being left vacant where the government encourages local people to plant fruit trees.

From observation, most people in the community lead a modest lifestyle. Though not having high monetary income, they seem to have enough resources for simple living within their environment. As they grow fruit and vegetables, they have enough food for living and some products for sale (except years when challenges like salt-water invasion occur). Generally speaking, community members seem to have good food security. Many senior retirees from urban employment have returned home to work on their gardens and grow fruit trees. People seem to be happy with a slower lifestyle with enough time to do what they prefer with their family and friends. There are many temples which people frequent for religious activities/festivals and for merit making. Temples are centers for community members' meetings, relaxation, and meditation or Dhamma practices. The community has identity built on being a local Buddhist culture, environment, and food production; members are proud of their green community. The green community also draws interest from many agencies trying to preserve the environment; thus, there have been numerous meetings calling community members to participate in related projects—active community members also often meet each other in such meetings or activities. There are also a number of community activities where people volunteer to be of service or represent their groups.

This chapter builds on Kittiprapas (2020) by further exploring other information and variables such as deriving more specific happiness from sustainable living associated with preserving the Bangkajao green-space, the relationship between the environment and inner happiness, and descriptive community information. As the community has preserved the green environment for numerous years and actively participates in environmentally concerned activities, I hypothesize that living within this green environment seems to be a main source of the community's happiness. Thus, within this chapter, I test the relationship between happiness and living in Bangkajao using mixed methods.

9.3 Method of Analysis

This chapter combines qualitative and quantitative techniques and analysis. Specifically, I draw on prior participatory action research focused on water management in the Bangkajao community over the course of nearly three years from 2017 to 2020. With respect to qualitative research, I conducted observational studies, interviewed key informants who are natural leaders and local residents in the community, and facilitated conversations during various workshops and meetings.

Building on the background qualitative information, I also developed and administered a quantitative survey (N = 490) on happiness-related issues across six sub-districts in the Bangkajao community, through late 2018 to early 2019. Questions centered on both objective and subjective variables of relevant happiness domains (i.e., income, employment, physical and mental health, social (community) and family relationships, environment, as well as subjective aspects of inner happiness (mind and wisdom-based), time used, community related factors such as governance, culture or local wisdom, and some open-ended questions. Given that many of the indicators (inner happiness proxies) are subjective, respondents were asked to self-evaluate their own happiness. Target participants included those who have lived in the community for a long period, through generations, or those who were born there. They were asked to participate through random sampling in different types of locations: houses, temples, parks, street corners, food shops and markets. In total, we obtained 490 responses from all sub-districts across Bangkajao, which goes beyond the recommended 400 samples in the Yamane formula (1967).

With the data collected, the chapter empirically tests the relationship between happiness and some proxy indicators of inner happiness at the mind and wisdom levels from the BSD framework (Kittiprapas, 2018a). I also test other domains affecting happiness on the physical level often demonstrated as significant in happiness studies. Domains are included from economics like income and job, and non-income measures such as health, relationships, time, and environment (Bataineh, 2019; Clark et al., 2008; Easterlin, 1974; Kittiprapas et al., 2007; Krekel & Mackerron, 2020; Kusago, 2007; Layard, 2005; MacKerron & Mourato, 2013; Reis & Gable, 2003). In addition to testing the associations between these domains and overall happiness, this chapter also tests more specifically the relationship between inner happiness indicators and the happiness derived from preservation of the green-space. This chapter also aims to investigate the relationship between inner happiness and environmental quality reflecting sustainable development drawing on the case study of Bangkajao.

From the survey data, a levels of happiness scale was used as the dependent variable Y while socioeconomic and inner happiness indicators were selected as independent variables Xi in the empirical model as: $y = \alpha + \beta iXi + \mu$. Because all dependent variables, Y, were in ordinal scales, various estimations (shown in respective Tables in the results section) were run in an ordered logit model using the STATA statistical tool.

Several estimations applying the above model are provided in the result section. The first estimation was to test overall subjective well-being (Y, ranking from 0 to 10)[2] and inner happiness indicators (Xi) such as giving, management of suffering, feeling one has enough, and levels of material wants beyond basic needs. The second estimation uses the interaction term of some inner happiness indicators.

Then, to test how inner happiness indicators relate to specific types of happiness from environmental sustainability in this community, another variable of happiness from preserving environment sustainably (ranked in 5 ordinal scales) was used as Y with another set of independent variables of inner happiness indicators such as giving/helping, feeling one has enough (with existing materials), want (in materials beyond basic needs), and suffering management. In addition, rankings of environmental management[3] was also added as another independent variable to test how it relates to happiness from preserving the green-space.

Next, the paper aimed to explore factors effecting environmental quality, by using the level of the quality of the environment ranked by the respondents in 5 ordinal scales as the dependent variable (Y). It was paired with a set of independent variables which are environmental management, community factors, such as good governance of local government, local culture and wisdom, as well as inner happiness like feeling one has enough and happiness from environmental preservation.

Finally, the chapter tested the significance of key socioeconomic indicators such as income, job, and non-income, i.e., health, family and community relationships, time, environment, and inner happiness indicators such as giving, feeling one has enough, and suffering management, to overall happiness as Y (ranking from 0 to 10). Some descriptive characteristics are provided in the following, before moving to the empirical analysis section.

9.4 Results and Discussions

Before going to empirical results, some key socioeconomics data from the survey highlighting characteristics of Bangkajao living are presented below.

9.4.1 Results from Descriptive Survey Data

Selected socioeconomic variables are shown (Figs. 9.2, 9.3, 9.4, 9.5, 9.6, 9.7, 9.8, 9.9, 9.10 and 9.11) with some supplemental data and results of the descriptive analysis.

---

[2] Ranking from low to higher level of happiness which 0 refer to not happy at all and 10 refer to extremely happy.

[3] In this model, the environmental management is used as a proxy environment variable to reflect how well the community manages to preserve environment. This variable is used instead of environmental

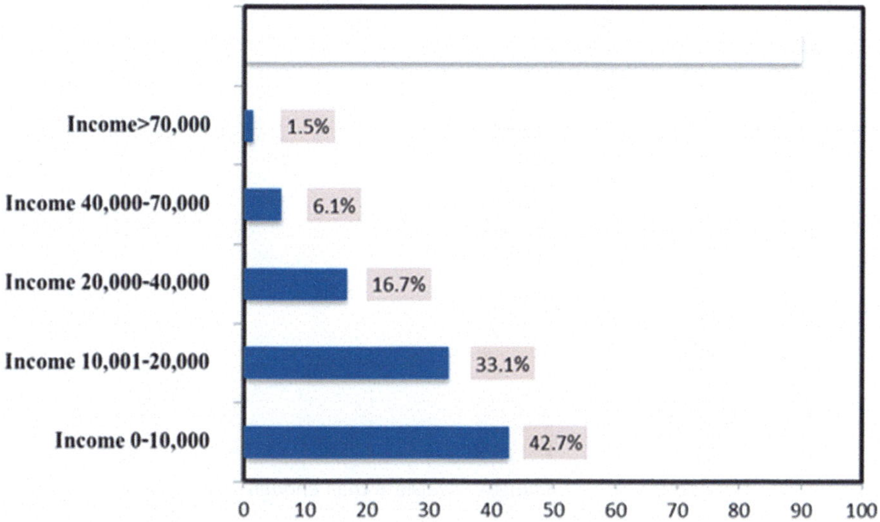

Fig. 9.2 Levels of income (1 Baht = 0.30 USD)

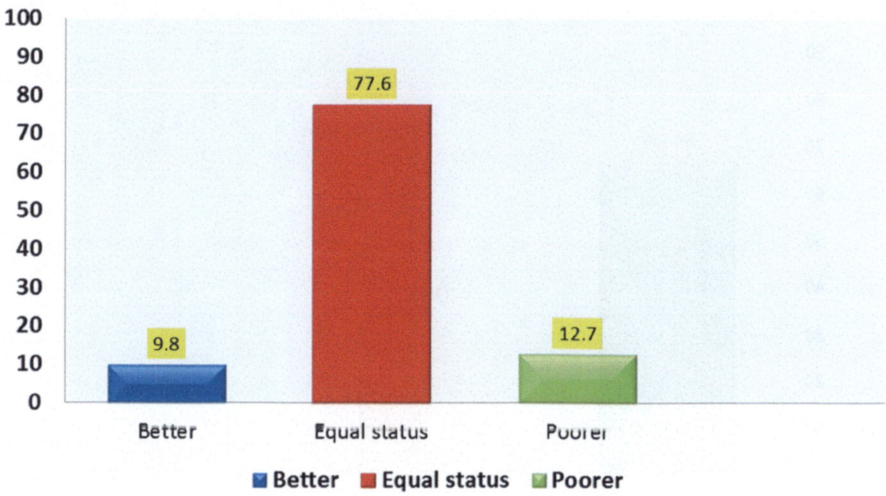

Fig. 9.3 Perception in relative income (to their community members)

Compared to bigger urban cities across the river, people in Bangkajao have very low income with many (43%) having less than 10,000 Baht monthly (~300 USD),[4] as

quality because, in this estimation, the dependent variable is happiness derived from preserving the green-space which may correlate to environment quality.

[4] With exchange rate approximately 33 Baht per 1 US $. Exchange rates in 2018 ranging between 31.127 and 33.464 with the average of 32.325 Baht to USD.

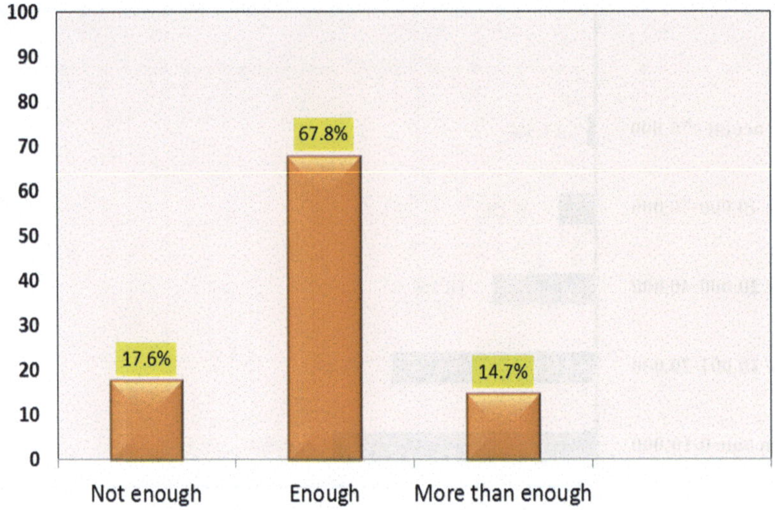

Fig. 9.4 Family economic management (family income to cover expenditures)

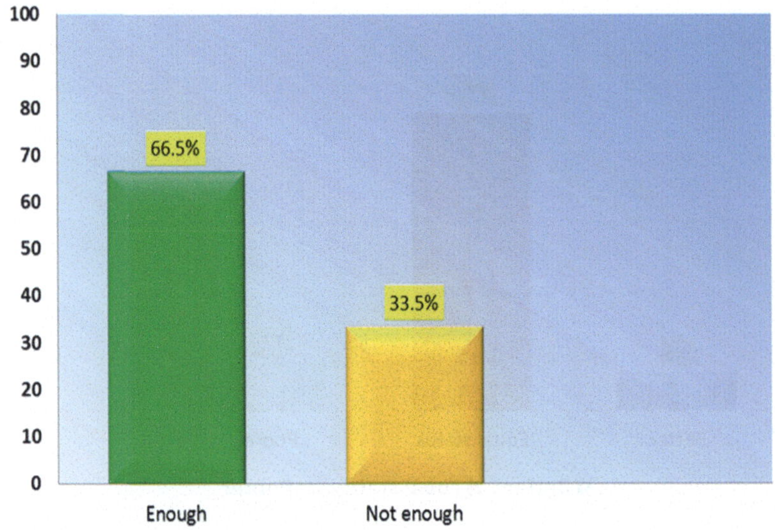

Fig. 9.5 Feeling one has enough in materials

shown in Fig. 9.2. However, the majority (about 78%) feel they have equal economic status to their neighbors or community members and about 10% feel better-off, as shown in Fig. 9.3. The result reflects their perception in relative income, indicating most feel equal or good. Thus, the low level of monetary income does not seem to affect happiness as the majority do not feel poorer. The results also seem to

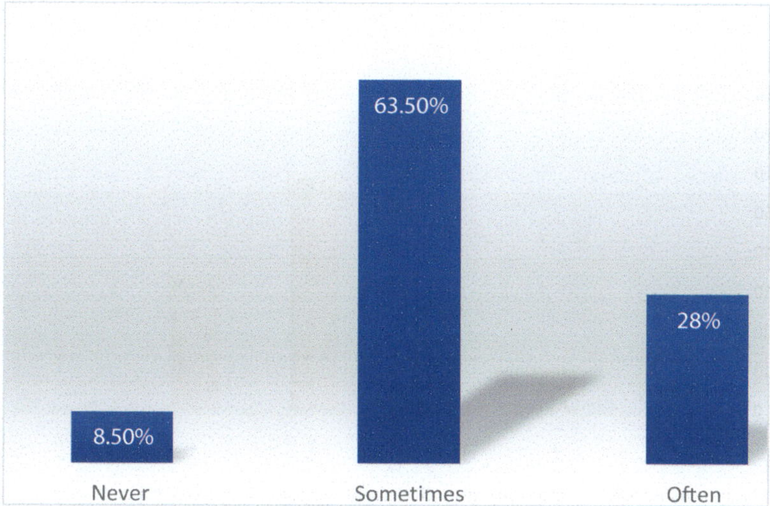

Fig. 9.6 Frequency of donation or volunteer work in a month

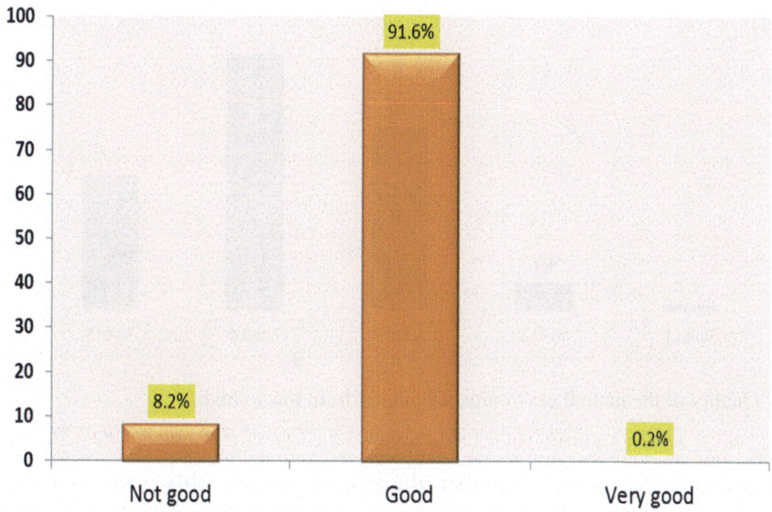

Fig. 9.7 Relationship within family

demonstrate that social comparison (which can lead to negative impact on happiness) has little effect for this community.

Concerning household income of the community, although average absolute income is not high (as shown in Fig. 9.2), very few households report having debts. Figure 9.4 indicates household economic situations and whether participants feel they have enough income for expenses.

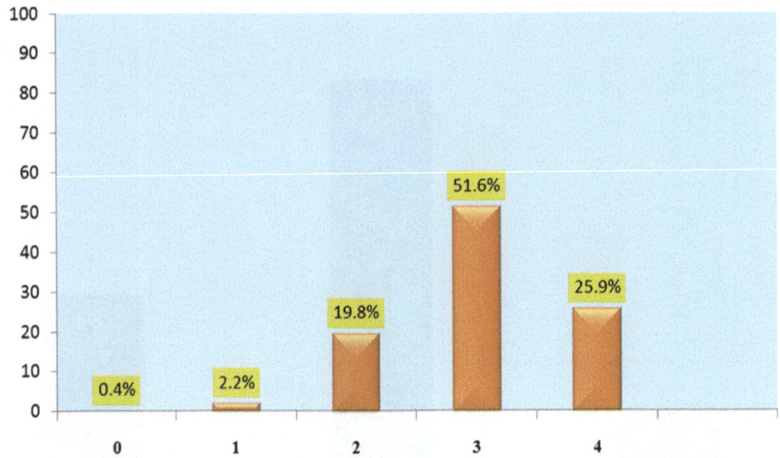

Fig. 9.8 Relationship with community members (ranking from low to high levels)

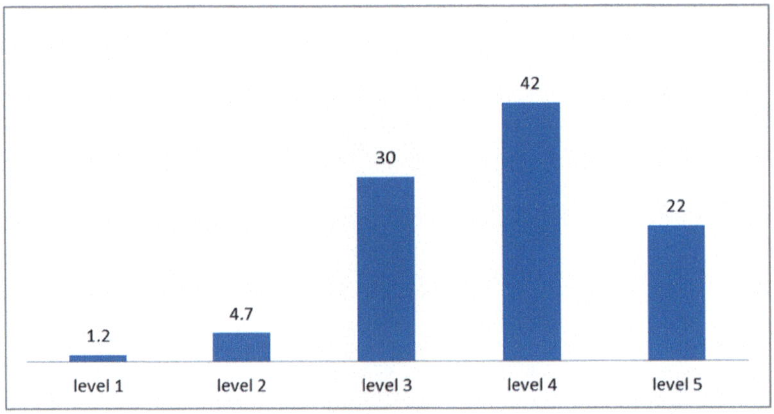

Fig. 9.9 Quality of the natural environment (ranking from low to high)

Respondents were asked whether their levels of monthly income can cover expenses or not. The results (Fig. 9.4) demonstrate that the majority of households (68%) have enough family income for living expenses and some of them (about 15%) are able to save some income, while the rest (~18%) experience financial problems. Thus, the majority are able to be self-reliant with respect to economics. The result is an example of the sufficiency economy concept (Wibulswasdi et al., 2011) as a practical model of simple living for happiness. From observation, although the majority of community members may have low income, they have in-kind support from their agriculture and environment for sufficient living or at least basic food

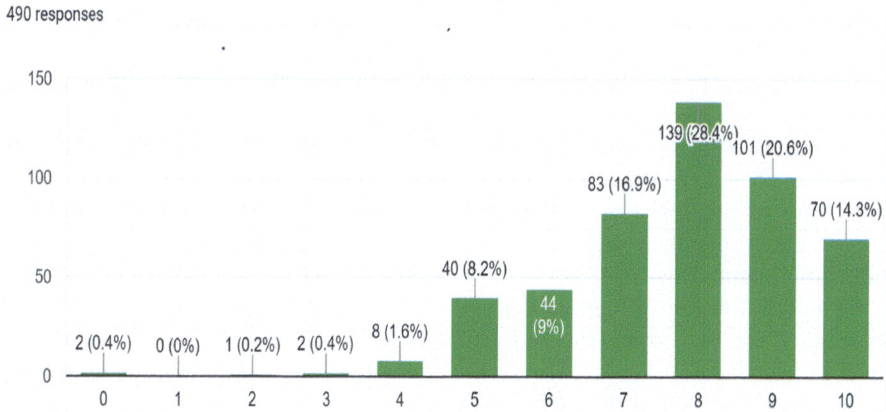

Fig. 9.10 Overall community happiness levels (0= not happy at all, 10= extremely happy)

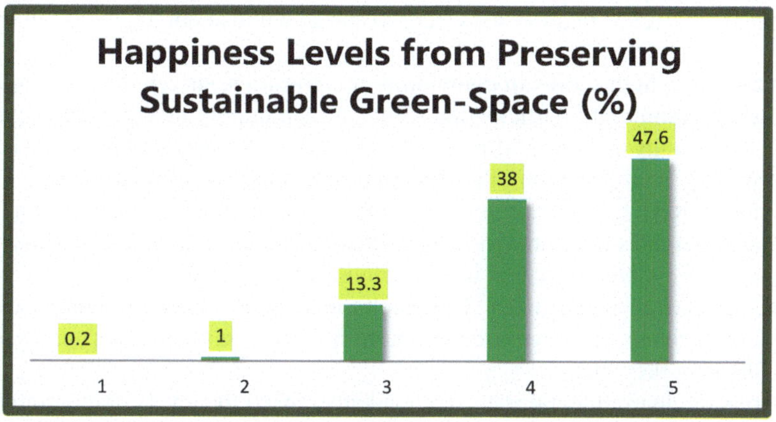

Fig. 9.11 Happiness levels from sustainably preserving green-space (0= not happy from preserving, 5= extremely happy)

security. As long as they are fulfilled by basic needs, they can be happy with moderation and simple living, in harmony with nature, according to Buddhist economics (Kittiprapas, 2017).

To check whether they feel contentment with what they have, the questionnaire also asked whether they feel they have enough with their current level of materials. Figure 9.5 shows that most feel they have enough indicating satisfaction with their material wealth.

Figure 9.5 reflects that nearly two thirds of residents find inner happiness from contentment with what they have. The result reflects little negative effects from aspiration and adaptation on the happiness of the community.

Figure 9.6 presents the frequency per month of those who donate per month or volunteer to help others, representing a giving variable.

After being asked how often they donate or help others within a month, the majority showed they exhibit high levels of giving to others in terms of frequency of donation and volunteer work. More than 90% donate or volunteer to help others and as many as 28% do so often.

In terms of relationships within the family, Fig. 9.7 indicates that the majority of community members have good relationships with their families and other community members.

As reported in Fig. 9.7, more than 90% of respondents have good family relationships. From the survey, the majority also feel proud of their family members. Therefore, having a strong family with good relationships is likely to be an important happiness determinant of this community.

Respondents were asked to rate the strength of their community relationships on a scale from 0 (low relationship) to 5 (high quality relationship).

The results (Fig. 9.8) shows high levels of relationship quality, with only 2.6% indicating low level of relationships (level 0 and 1) while about 87.5% are in level 3 and 4.

In addition to high scores in community relationships, survey data also indicate high rates of community participation—84% of community members have participated in community activities. Most also feel they can rely on neighbors when having problems. This is in line with other research indicating social relationships as key to happiness (Reis & Gable, 2003). In addition, survey data also shows that most community members have time to spend with their family, experience personal pleasures, and express creativity and useful activities that they enjoy doing (only 16% reported not having much time). The results indicate that the community, experiencing simple living, can have good relationships amongst themselves and available time to do what they like.

In terms of environmental issue, respondents ranked the levels of environmental quality as indicated in Fig. 9.9.

Specifically, respondents were asked to rate the quality of the natural environment in their community from very poor (level 1) to excellent (level 5). The most frequent response was level 4 (42%), with about 94% of respondents selecting levels 3–5, indicating a positive perception about the quality of the community's natural environment.

The distribution of overall community happiness levels is shown in Fig. 9.10. Respondents were asked to rate their happiness on a score of 0–10, with 0 being not happy at all and 10 being extremely happy.

From Fig. 9.10, the highest frequency for happiness levels (from 0 to 10) is level 8, followed by level 9, 7, and 10, respectively. The average score for happiness of the community is 8.5, indicating a high average happiness level in a relatively low-income community within a pretty good natural environment.

Respondents were also asked to self-evaluate their happiness levels derived from preserving their green-space community, as it has a history of fighting the demand of previous governments to give up their land. Specifically, residents were asked what level of happiness (1 = lowest and 5 = highest) was derived from the sustainability of their green space. Nearly half the respondents answered chose level 5 (Fig. 9.11).

As indicated in Fig. 9.11, nearly 99% of the respondents chose levels 3–5 with respect to the happiness derived from preserving their sustainable green space. This indicates a strong perceived relationship between happiness and sustainable development and may explain why this community has preserved their green environment for a long time. The result also may indicate how sustainable environment preservation has highly affected the community's happiness for decades, even though the average community income is relatively low.

The above results demonstrate characteristics that are unique to the people in the community. These data support the concept of inner happiness, such as the feeling of contentment, rather than wanting greater and greater monetary income and material wealth. However, factors contributing to the respondents' happiness are further investigated in the empirical estimations below.

9.4.2 Empirical Tests

Before testing key socio-economic factors affecting overall happiness, the chapter firstly focuses on the role of inner happiness variables (mind and wisdom based) and environmental indicators as independent variables Xi; namely; the frequency of giving or helping others, levels of materials desires, feeling one has enough or contentment, and the ability to manage suffering, as well as environmental quality.

With a statistical confidence level of 95%, Table 9.1 shows that feeling one has enough (with respect to materials) has the largest significance (~0.75), followed by environmental quality (~0.67). The strongest effect is feeling one has enough in materials on overall happiness. In other words, it reflects a low level of material aspiration in the community as they feel contentment with what they have. The result is consistent with other works mentioning the negative effects of material aspiration (Di Tellar et al., 2005; Hussian & Heshmat, 2010; Inglehart & Rabier, 1986).

The results indicate that inner happiness indicators such as feeling one has enough and suffering management are significant and positively influence happiness, while giving and level of wants beyond basic needs are not significant. However, survey data indicate that giving has a very high frequency over that of not giving; thus, the

Table 9.1 Empirical tests of overall happiness with environmental quality and inner happiness indicators n = 490

Happiness	Coefficient	P > Z
Environmental quality	0.669295	0.000
Giving	0.3324683	0.282
Suffering management	0.5271632	0.000
Feeling one has enough	0.7474979	0.000
Level of wants	−0.0898805	0.219

Table 9.2 Empirical tests of overall happiness with environment quality and inner happiness indicators (with interaction term)

Happiness	Coef	P > Z
Environmental quality	0.6802536	0.000
Giving/feeling one has enough	0.5969852	0.002
Suffering management	0.5495491	0.000
Wants	−0.116788	0.105

insignificance may be due to a general worldview and practice of giving within this particular group. Thus, the following estimation tests the interaction term of giving and feeling one has enough, rather than examining them separately. The results in Table 9.2 show that those who have the characteristics of feeling one has enough in materials and associated with the character of giving to or helping other people significantly contributes to happiness, as presented in Table 9.2.

Table 9.2 indicates that the interaction terms of feeling one has enough and giving, reflecting that the feeling one has enough and being willing to give/help others, significantly contributes to happiness. Thus, people who feel content with what they have or those feeling they have enough can give or share things or time to others. The resulting inner happiness from being content and sharing helps to increase their overall happiness (0.60). In addition, the result (0.55) of the wisdom based indicator suffering management shows that a high level of skill in suffering management can lead to higher levels of happiness. Although want is not significant at a 95% confidence interval, it is significant at 90% confidence interval with a negative relationship to happiness, implying the less wants in materials (beyond basic needs), the higher level of happiness. Consistent with the other estimation, environmental quality is largely significant to happiness and, for this estimation, has the largest coefficient (0.68). This is consistent with other studies indicating the importance of environment to happiness (Kittiprapas, 2020; Krekel & MacKerron, 2020; MacKerron & Mourato, 2013).

The results of all inner happiness indicators tend to support the BSD argument that overall happiness can be increased through and with enhanced inner happiness. Next, the paper tests how inner happiness indicators relate to specific types of happiness from environmental sustainability in the Bangkajao community (Fig. 9.11) instead of overall happiness. Another set of independent variables of inner happiness indicators such as giving/helping, feelings of having enough, wants, and suffering management, in addition to management in environment are added as independent variables to test how they relate to happiness from preserving this green space (Table 9.3).

This case shows that all inner happiness indicators are statistically significant at 95%; namely, giving/helping, feeling one has enough, management of suffering, and material wants (with a negative effect). Giving and helping others is significant with the highest coefficient (0.65) to happiness from preserving the environment in this case. It might imply that sustainable environmental preservation matters to happiness, as this community is willing to give time/things and help others. Such inner happiness can lead to the increased happiness from sustainably preserving the environment. Next, the environmental management variable is also highly significant, reflecting

Table 9.3 Empirical tests of happiness from preserving green space and inner happiness n = 490

Happiness from Preserving Green Space	Coef	P > Z
Environmental management	0.53838	0.000
Giving/helping	0.6507969	0.038
Feeling one has enough	0.496913	0.024
Wants	−0.1612719	0.047
Suffering management	0.316833	0.04

that the better management of environment, the higher the happiness from preserving it. Consistent with other tables, feeling one has enough is positively significant, while want is negatively significant meaning the lower wants for materials leads to higher level of happiness from preserving the environment. These results confirm that inner happiness from feeling one has enough, lower wants in materials, as well as giving or helping others may lead to higher happiness from preserving the environment.

The provided estimations (Tables 9.2 and 9.3) show that inner happiness indicators relate positively to the overall levels of happiness in general and the specific happiness from preserving the environment in particular. Obviously, higher inner happiness from feelings of having enough, giving/helping, less material wants, in addition to environmental management, significantly lead to the increases in happiness from preserving the green space, as shown in Table 9.3.

Additional empirical tests were conducted to examine the relationship between inner happiness and environmental quality as well. The test aims to explore factors affecting environmental quality and examine the relationship with a set of independent community variables (i.e., environmental management, good governance of local government, local culture and wisdom, and inner happiness such as feeling one has enough and happiness from environment preservation). The objective is to test whether environmental quality relates to inner happiness and those community factors (Table 9.4).

The results show that inner happiness (represented by feelings of having enough or feelings of contentment) significantly contributes to environmental quality (with 0.44 coefficient). Other significant factors are good environmental management (with the largest coefficient of 0.99), happiness from preserving the green space (0.63), and the strength of local culture/wisdom. However, local good governance (by ranking how well the local government organizations work) is not significant to environmental

Table 9.4 Empirical test of environmental quality with community variables and inner happiness n = 490

Environmental quality	Coef	P > Z
Environmental management	0.9928929	0.000
Green happiness	0.6345244	0.000
Local governance	−0.1566068	0.307
Local culture/wisdom	0.2716692	0.003
Feeling one has enough	0.4365301	0.021

Table 9.5 Happiness effects from socioeconomics and inner happiness indicators n = 490

Happy	Coef	P > Z
Income	0.0233719	0.790
Perception in relative income	0.592907	0.002
Job	−0.3726596	0.077
Time	0.5001345	0.012
Health	0.539598	0.000
Family relationship	0.6860511	0.030
Community relationship	0.4435657	0.000
Environment quality	0.5239902	0.000
Giving	−0.1557799	0.628
Feeling one has enough	0.7805683	0.000
Suffering management	0.2898464	0.009

quality. Interestingly, environmental quality seems to be largely affected by local people's inner happiness such as the happiness derived from preserving this green-space area, feeling one has enough in materials, and local culture/wisdom.

In addition, empirical results in Table 9.5 provide an overall picture of factors affecting the happiness of this community, which include key socioeconomic indicators from happiness theories (income, relative income, job, time, health, family and community relationships, environment, as well as inner happiness form BSD (Kittiprapas, 2018a). The Proxy of inner happiness indicators used in this estimation include giving, feeling one has enough, and suffering management.

From Table 9.5, it is interesting to see that, while monetary income is not significant, the perception in relative income/economic status is evidently significant. Thus, the results show that the perception of economic status relative to their neighbor (or subjective aspect of income), is more important than objective incomes (or the monetary/absolute income) also consistent with other happiness studies (Knight & Song, 2006). From descriptive statistics in Figs. 9.1 and 9.2, although they have low monetary income, the majority of respondents do not feel poorer than others, so their perception in relative income is pretty good. The low effect of social comparison may allow them to be happy; their family and community relationship also boost happiness levels. The community's quality of environment also raises their happiness on a large scale, in addition to good health and available time to complete useful activities (Alesina, et al., 2006; Bataineh, 2019; Hussien & Heshmat, 2011; Layard, 2005; Sabatini, 2014). Relationships within both the family and the community are very important as both indicators are significant (Clark et al., 2008; Helliwell, 2003; Helliwell, 2006; Liping, 2001; Mendes et al., 2007; Reis & Gable, 2003; Sarracino, 2007). Inner happiness indicators of having enough in material resources and suffering management are also significantly positive to life satisfaction or happiness (Kittiprapas, 2020).

Considering Table 9.5, the feeling one has enough variable has the strongest effect, followed by family relationships, environment quality, perception in relative income, health, time balance, and community relationships, which are among the strongest effecting variables, respectively. Additionally, suffering management is significant to happiness. Although having a job usually makes people happy, the job variable here is not significant, which may be due to the fact that many community members have jobs in the Bangkok and Samutprakarn areas where they have to travel daily to work. More, that type of work may not make them happy, in addition to the uncertainty of agriculture facing salt water invasion. The contradiction of job satisfaction may also have resulted in the insignificance of the job variable. Also, giving is not significant in this general case, but results from Tables 9.2 and 9.3 show that when it interacts with feeling one has enough or in relation to happiness from preserving the green space, giving has significantly positive results. However, this table confirms the significance of inner happiness at the wisdom level—to know what is moderate or when one has enough supports a real quality of life and the ability to manage one's suffering.

Empirical tests for factors contributing to happiness discussed in this paper are in line with the above descriptive statistics and also consistent with other results (with different models) found in Kittiprapas (2020). The results also indicate that the community members are happy from feeling they have enough in materials as well as environmental quality which is significant across all classified groups in this community. Similarly, absolute income is not significant to happiness, while relative income does matter.

The community study shows the importance of subjective aspects of happiness from a green and low-income community with a unique landscape and history/culture. Inner happiness positively effects the community's environmental quality and also contributes to overall happiness. The community members have high levels of happiness from preserving the community green-space and inner happiness is also relevant, supported by the empirical tests. These results imply that, with this kind of simple living, economic income may not be the ultimate aim but other factors (such as inner happiness, environment, relationship, health and time balance) matter for their happiness. The sufficiency in materials and inner happiness can allow residents to live happily in harmony with green-space environments while also working to maintain them. Life satisfaction comes from subjective aspects such as the feeling that they are not in poorer economic status, having good relationships, pride with family members and local wisdom/culture as well as good community environments. With moderate and simple lifestyles, they have enough time to do what they like to do and help others as well as having good psychological and physical health. Qualitative research also supports these characters of the community's happiness. They have satisfaction with the environment surrounded by greenspace which they have preserved for a long time, while other areas on the other side of the river (Bangkok and Samutprakarn) have become concreted urban spaces.

All supporting factors of happiness in the community reinforce one other. A good environment makes residents healthy and happy. They are aware that their green space is unique and useful for their quality of life; it produces fresh air and reduces pollution and global warming, a fact they are proud of along with their willingness to help others in order to keep the environment sustainable. Good relationships within the family and community members also makes them happy. Friendship and compassion for others and nature can also lead to happiness. Simple living also allows them more time to spend on preferable activities, i.e. relationships and environmental conservation, in addition to going to temples for religious occasions and mindfulness practice. Living in moderation with time balance can also benefit health and generate inner happiness. In turn, as one estimation indicates, inner happiness highly leads to the enhanced environmental quality, leading toward sustainable development.

Inner happiness, particularly the feeling of having enough, is crucial for determining the happiness of this green-space community, as also evident in Kittiprapas (2020). Considering the environmental dimension, the paper found that the environment is positively significant to happiness of all sub-groups in the community regardless of income levels, occupations, or age group. The result shows that environmental quality is a common factor affecting the happiness of the community. Different kinds of estimations in this chapter also confirm the significance of inner happiness and environmental preservation to this community happiness.

The main empirical finding in this article indicates that inner happiness significantly relates to happiness as well as environment of this community. The feeling of having enough in materials is essential for this community's happiness and for a long history of preserving the environment. This is a key influential factor for happiness of people in this community who have lived within the green area for centuries. If people keep on increasing the demand for materials and do not feel as though they have enough or are not satisfied with what they have, more and more natural resources will be used and green areas will have been lost. Aspiration for more materials will lead to adverse impacts on the environment. The green space in the studied community would quickly decline if the residents change their lifestyles for increasing materialism. As mentioned in the BSD approach, people can be happy with modest use of resources and consumption. In fact, happiness from having enough in material wants also reflects inner happiness at the wisdom level, in that they know what levels are moderate or appropriate, or what is *real* happiness and what causes suffering. Therefore, happiness comes from feeling one has enough, rather than a continuous search for more materials to meet desires that never ends and brings suffering. This case study also suggests that economics of moderation may lead to low income inequality and low social comparison, which can increase community happiness. It is essential that most people feel that they have similar economic rank as others in the community and do not feel relatively poor, although their income level is relatively low compared to other cities.

The limitation of this study, however, includes the survey being completed only during the daytime. It likely excluded some workers who commute to work in other areas outside the community. More, the possible changes in demographic factors in the near future, with new residents and tourists as well as outside organizations involved in the community, and the influences of globalization and urbanization may permanently change the residents' life styles and occupations.

Inner happiness, as evidenced by this study, can be in line with sustainable development. This kind of happiness can be generated inside one's own mind, rather than from meeting physical wants from the outside that leads to more acquisition and material accumulation in conflict with a sustainable development pathway. Inner happiness with less acquisition and the elimination of unnecessary wants does not exploit natural resources and other living beings; thus, it should be promoted to make sustainable development become a reality. In other words, striving for inner happiness can fill the gap of *how to* achieve *sustainable development* which is missing in the Bruntland definition. The BSD is a practical direction towards sustainability and can promote a sustainable future with greater levels of happiness.

9.5 Conclusion

The chapter shares details of a study conducted in a green-space community in Thailand aiming to test factors affecting overall happiness as well as happiness derived from preserving their environment. In addition to confirming the significant role of inner happiness on overall happiness, the chapter also investigated the relationship between inner happiness and the environment. In sum, the empirical findings discussed in this paper confirm that various indicators for inner happiness with the moderation in material living significantly related to happiness of the community members and environmental preservation. Statistical tests show the relevance of inner happiness at both the mind and wisdom levels (i.e. generosity, feeling one has enough, less want, and the ability to manage suffering). Social comparison (a significant factor of unhappiness in many studies) was low in this community, where monetary income does not matter as long as most of the people feel that they have similar economic status/rank alongside good financial management within households. The sufficiency in materials and management of expenditures within limits also reduces the negative effect of aspiration/adaptation on happiness. Thus, residents can manage their situation to live happily through simple living, which is supported by various descriptive statistics from the survey. The community members here seem to have a slow life with enough time to do preferable things and have a good family relationship; they are proud of their family members, their own culture and local wisdom, environmental quality and management as well as their long history of preserving it.

Ultimately, the case of Bangkajao community's happiness supports the concept of Buddhist sustainable development (BSD) where inner happiness is key to a sustainable future. Such happiness comes from simplicity and moderation where a community maintains their environmental sustainably. Thus, inner happiness, especially feeling one has enough or experiencing sufficiency in materials, should be promoted as a pathway for sustainable development. Overall, the results of this case study contribute to happiness research with additional effects of inner happiness from the mind and wisdom levels, apart from other well-known factors (health, economics, and family/social relationship, etc.). In addition, the chapter shows the link between environmental sustainability and inner happiness. The findings from this case study can provide a direction for protecting environmental areas and sustainable development which can be used across cultures and regions.

References

Alesina, A., Glaeser, E. L., & Sacerdote, B. (2006). Work and leisure in the U.S. and Europe: Why so different? In M. Gertler & K. Rogoff (Eds.), *NBER macroeconomics annual 2005* (pp. 1–64). MIT Press.

Bataineh, K. (2019). Impact of work-life balance, happiness at work, on employee performance. *International Business Research, 12*(2), 99–112.

Bouchet, C. M. (2013, June 14). Escaping to "Bangkok's green lung." *New York Times*. https://www.nytimes.com/2013/06/16/travel/escaping-to-bangkoks-green-lung.html

Clark, A. E., Frijters, P., & Shields, M. A. (2008). Relative income, happiness and utility: An explanation for the Easterlin paradox and other puzzles. *Journal of Economic Literature, 46*(1), 95–144.

Easterlin, R. A. (1974). Does economic growth improve the human lot? Some empirical evidence. In P. A. David, & M. W. Reder (Eds.), *Nations and households in economic growth: Essays in honor of Moses Abramowitz* (pp. 89–125). Academic Press. https://doi.org/10.1016/B978-0-12-205050-3.50008-7

Helliwell, J. F. (2003). How's life? combining individual and national variables to explain subjective well-being. *Economic Modelling, 20*, 331–360.

Helliwell, J. F. (2006). Well-being, social capital, and public policy: What's new? *Economic Journal, 116*(510), 34–45.

Hussian, S., & Heshmat, Y. (2010). Determinants of happiness and life satisfaction in Egypt (Working paper No.13). Information and Decision Support Center-Working Paper Series (IDSC-WPS).

Inglehart, R., & Rabier, J. R. (1986). Aspirations adapt to situations—But why are the Belgians much happier than the French? A cross-cultural analysis of the subjective quality of life. In F. M. Andrews (Ed.), *Research on the quality of life* (pp. 1–56). Institute for Social Research University of Michigan.

Kittiprapas, S. (2017). *Buddhist economics: A holistic development approach for true well-being* [e-book]. International Research Associates for Happy Societies (IRAH).

Kittiprapas, S. (2018a). *Buddhist sustainable development: A new approach in happiness and development* [e-book]. Thammasat University Press.

Kittiprapas, S. (2018b). What really determine happiness of different groups: A case study of Bangkok. *Journal of Business and Economics, 9*(5), 318–329.

Kittiprapas, S. (2020). Happiness determinants in a Buddhist community: Where inner happiness matters. *Thammasat Review of Economic and Social Policy, 6*(1), 84–135.

Kittiprapas, S., Sawangfa, O., Fisher, C., Powdthavee, N., & Nitnitiphrut, K. (2007). Happiness: As a new paradigm for development and public policies. *Chulalongkorn Journal of Economics, 19*(3), 242–281. http://www.econ.chula.ac.th/public/publication/journal/2007/cje190303.pdf

Knight, J., & Song, L. (2006). Subjective well-being and its determinants in rural China. University of Nottingham.

Krekel, C. & Mackerron, G. (2020). How environmental quality affects our happiness. In J. Helliwell, R. Layard, J. D. Sachs, & J. E. De Neve (Eds.), *World Happiness Report* (pp. 95–112).

Kusago, T. (2007). Rethinking of economic growth and life satisfaction in Post-WWII Japan—A fresh approach. *Social Indicators Research, 81*(1), 1–24.

Layard, R. (2005). *Happiness: Lessons from a new science*. Penguin Books.

Liping, X. (2001). The relationship between happiness and social support. *Journal of Chinese Psychology: Acta Psychologian Sinica, 33*(05), 59–64.

MacKerron, G., & Mourato, S. (2013). Happiness is greater in natural environments. *Global Environmental Change, 23*(5), 992–1000. https://doi.org/10.1016/j.gloenvcha.2013.03.010

Marshall, A. (2006, May 15) Best urban oasis. *Time Magazine Asia*. http://content.time.com/time/subscriber/article/0,33009,1194117,00.html

Mendes, J., Guerreiro, M. M., & Vieira, G. (2007, July). Students' happiness and wellbeing at the University of Algarve, Portugal. In *United Nations Conference Center (UNCC) Bangkok, Thailand*.
Reis, H. T., & Gable, S. L. (2003). Toward a positive psychology of relationships. In C. L. Keyes & J. Haidt (Eds.), *Flourishing: The positive person and the good life* (pp. 129–159). American Psychological Association.
Sabatini, F. (2014). The relationship between happiness and health: Evidence from Italy. *Social Science & Medicine, 114*, 178–187.
Sarracino, F. (2007, July 18–19). Subjective well-being in low income countries: Personal, relational, and social capital components (paper). In *International Conference on Happiness and Public Policy, Bangkok. Public Policy Development Office and UNESCAP.*
Time (2006, May 22). *The Best of Asia, 167*(21).
The World Commission on Environment and Development (The Brundtland Commission). (1987). *Our common future*. Oxford University Press.
Welsch, H. (2002). Preferences over prosperity and pollution: Environmental valuation based on happiness surveys. *KYKLOS, 55*(4), 473–494. https://doi.org/10.1111/1467-6435.00198
Wibulswasdi, C., Piboolsravut, P., & Pootrakool, K. (2011). *Sufficiency economy philosophy and development*. Bureau of the Crown Property.
Yamane, T. (1967). *Statistics: An introductory analysis-3* (2nd edn.). Harper and Row.

Chapter 10
How Sustainable is Mobility in Cities Branded the Happiest?

Alshimaa Aboelmakarem Farag

Abstract In recent years, the interest in measuring happiness has increased. Numerous initiatives measure and market happiness interpreted through metrics such as housing affordability, unemployment rates, and safety. However, they have not considered sustainable mobility as a possible indicator of communities' happiness levels. This chapter shares the results of a mixed methodology aiming to link sustainable mobility to the happiness levels of urban people via three objectives. The first objective is to identify the key role of sustainable mobility in improving people's mood as it has a significant impact on quality of life and public health; consequently, it can also contribute to making people happier. The second objective is to investigate sustainable mobility's weighting factor within three international happiness indexes, the *World Happiness Report*, the *Happy Planet Index*, and the *Better Life Index*, as standards of measurement for people's happiness, branding the happiest cities to the global community. The third objective is to emphasize the sustainable mobility performance in those cities officially branded as the happiest in the world. This chapter concludes with happiness as a possible element that can be embedded more deeply in the discourse of promoting sustainable mobility.

Keywords Active travel · Happiness index · Sustainable transport · Travel behavior · City branding

10.1 Introduction

Happiness is a term with a broad spectrum of reflections, many of which have raised debates throughout history; humanity's interest in happiness is not a recent phenomenon. Even today, happiness is a top endeavor of humankind as people, communities, and organizations have been searching for methods to achieve and measure happiness in various life domains. Organizations and many countries worldwide have been working to measure happiness and market the happiest cities to the

A. A. Farag (✉)
Zagazig University, Zagazig, Egypt
e-mail: AATantawi@eng.zu.edu.eg

S. Cloutier et al. (eds.), *Linking Sustainability and Happiness*, Community Quality-of-Life and Well-Being,
https://doi.org/10.1007/978-3-030-89559-4_10

global community. The trend has resulted in branding a group of cities as those surpassing others in levels of happiness, advertising their people as some of the happiest on the globe. In line with this trend, happiness standards are becoming a top priority. Organizations like the World Bank and United Nations have supported happiness measurement efforts by initiating a series of metrics to measure happiness levels. The World Happiness Report, the Happy Planet Index, and the Better Life Index are global initiatives that aim to measure cities' happiness levels. The goal to pursue happiness has resulted in research exploring the term's definitions, dimensions, and measures. Happiness, as a term, has been used interchangeably with many other concepts, such as wellbeing and life satisfaction. In order to avoid a mix-up between happiness and other terms, happiness and subjective wellbeing are used as synonyms for the purpose of this chapter (Farag et al., 2019).

In the pursuit of happiness, many scholars tried to link the concepts of happiness and sustainability within and across different scales and settings. For example, Cloutier et al. (2014) emphasized the significance of considering happiness when measuring urban sustainability. Similar studies concluded that cities with higher reported scores in sustainability indices might offer their residents more significant opportunities for happiness (Cloutier & Pfeiffer, 2015; Leyden et al., 2011; Pfeiffer & Cloutier, 2016; Veenhoven, 1993; White et al., 2013, 2017; Zidansek, 2007). However, more evidence can prove happiness as an essential component of sustainability indexes, leading to future studies that might assess the influence of specific subsystems of sustainable development on happiness, also providing some insights into causality (Berejnoi et al., 2019; Cloutier et al., 2014; Leyden, et al., 2011; O'Brien, 2008; Schimmel, 2009).

Sustainable mobility is important to consider in achieving urban sustainability. The term *sustainable mobility* was launched in 1992 by the EU as part of the *Green Paper*. The intention was to present sustainable mobility as a strategy that might enable transport to fulfill its economic and social role while containing its harmful effects on the environment (Holden et al., 2019). The term is currently used to refer to the broad subject of mobility in that it is sustainable under the framework of social, economic, and environmental impacts. In other words, sustainable mobility refers to the available travel modes that meet people's present economic, social, and environmental needs while minimizing their undesirable impact on the economy, society, and the environment (Gallo & Marinelli, 2020). The most commonly used examples are walking, cycling, electric vehicles, scooters, vehicle sharing, metro, tram, and public transport. A requirement is that sustainable mobility modes function in a way that minimizes air pollution and noise, reduce CO_2 emissions compared to other modes, promote economic development, enable efficient mobility alternatives, and are affordable. Promoting sustainable mobility also requires significant actions to minimize the number of travel trips by reducing the need to travel, reducing trip lengths, and enhancing the efficiency of sustainable mobility services and infrastructure (Badawi & Farag, 2021; Oleśków-Szłapka et al., 2020).

In line with the importance of promoting sustainable mobility, the international community has shown serious commitment over the last several years, including many private and public organizations at different levels and scales. The United

Nations General Assembly set a Sustainable Development Goal (SDG11.2) to "make cities inclusive, safe, resilient and sustainable," which included:

> By 2030, provide access to safe, affordable, accessible and sustainable transport systems for all, improving road safety, notably by expanding public transport, with special attention to the needs of those in vulnerable situations, women, children, persons with disabilities and older persons (SDG11.2).

The UN Secretary General's high-level advisory group also framed sustainable mobility attributes to include safety, affordability, accessibility, efficiency, resilience, and carbon impacts (Vandycke et al., 2019).

Although public happiness is an underlying intention of many politicians and planners, the term happiness is rarely used in sustainable mobility discussions (O'Brien, 2008). More, the link between happiness and sustainable mobility and what underlies this relationship is unclear. It has not been studied extensively or proven consistently (Kim, 2018; O'Brien, 2008); yet, there are a few studies linking sustainable mobility and happiness. For example, Morris and Guerra (2015) suggest that transit riders' affective experience is improved through enhanced transit quality and affordability. Meanwhile, Cloutier et al. (2017) concluded that American cities with higher use of sustainable mobility modes and a dedicated commitment to improving sustainable mobility infrastructure have higher happiness levels among residents. This chapter is conducted with the intention of contributing to research indicating that sustainable mobility has a significant impact on quality of life, and public health, providing insights into sustainable mobility's potentials to make people happier. Below, I provide a review of relevant literature identifying the contributions sustainable mobility offers to happiness.

10.1.1 Sustainable Mobility and Happiness

This section aims to identify the key role of promoting sustainable mobility to improve environmental quality and its significant impact on collective quality of life and public health; consequently, it can contribute to making people happier. The chapter adopts the concept of happiness as the subjective construct of experience in which two features are addressed, the affective and cognitive (Farag et al., 2019), assuming that sustainable mobility plays a key role in improving each.

10.1.2 Sustainable Mobility and the Affective Happiness Experience

Sustainable mobility has an indirect impact on people's affective experience. Choosing mobility modes such as walking, cycling, and public transport embodies a behavior of physical activity, and many of those who use public transport rely on physical activity while they walk or bike to public transport stations. Several studies demonstrate associations between physical activity and happiness (Black et al., 2015; Steptoe, 2019; White et al., 2017). Practicing regular physical activity lowers exhaustion and stress levels, improves positive emotions, enjoyment, and interest (Hansson et al., 2011; Lee, & Shin, 2020), and releases serotonin, the *happiness hormone* (Morris & Guerra, 2015). Vice versa, a bidirectional association between physical inactivity and depression has been recognized for many years (Steptoe, 2019). Therefore, people who choose active travel frequently, such as walking and cycling, are often happier than their inactive peers (Gatersleben & Uzzell, 2007; Guell & Ogilvie, 2015; Martin et al., 2014; Olsson et al., 2013). For example, a study conducted on a secondary school shows that the adolescents who spent more than 15 min each day actively commuting to school had higher levels of subjective happiness and psychological wellbeing and lower levels of psychological distress than adolescents who spent 15 min or less per day (Ruiz-Ariza et al., 2015). Similarly, parents and children who used active travel modes on a school trip reported more positive emotions than other travelers (Ramanathan et al., 2014).

Engaging in regular physical activity through active travel is important to enhancing public health (Mytton et al., 2016; Rasciute & Downward, 2010), also making people happier (Richards et al., 2015). Research shows that happiness and health are related to some extent in that good health could positively impact a person's happiness and satisfaction. Similarly, being happy and in a positive mood can influence a person's health and immune system (Argyle, 1997; Willits & Crider, 1988), making people live longer (Deeg & Zonneveld, 1989; O'Brien, 2008; Veenhoven, 2008). Additionally, the positive affect of happiness and physical activity are recognized to reduce many causes of death (Hoen et al., 2013).

Morris and Guerra (2015), found that different modes of mobility may change the levels of happiness, sadness, joy, pain, and stress in travelers. They found that cyclists feel the most positive affect, followed by car passengers and drivers, unlike bus and train riders who experience negative emotions. It is a fact that driving private cars can positively influence people's happiness as it is a cultural symbol of freedom, power, speed, and social betterment, especially for seniors and the lowest income, and also when the trips are related to leisure rather than work (Duarte et al., 2010; Kim, 2018). Understanding such relationships would provide new insights on improving travel trip quality and transportation services to make the public transport experience as pleasant as other mobility modes and entice car drivers to leave their car for more sustainable choices. There may also be robust opportunities for additional empirical research on the affective impacts of mobility modes on people in different situations.

10.1.3 Sustainable Mobility and the Cognitive Happiness Experience

Sustainable mobility could contribute to people's cognitive happiness and life satisfaction. A wide spectrum of studies focuses on the various factors impacting the cognitive happiness levels of people who choose sustainable mobility for movement. For example, expecting pleasant events from a travel destination, such as meeting and catching up with friends, can impact people's feelings and provoke happiness, while experiencing crowding, traffic congestion, and increased travel time can cause unpleasant feelings (Leyden et al., 2011; Martin et al., 2014; O'Brien, 2008). Additionally, being unsafe or exposed to crime and high noise levels in the travel trip can undermine people's happiness (Forsyth, 2015; Morris, 2015; Morris & Guerra, 2015; Olsson et al., 2013). Living in a built environment that prompts sustainable mobility also is a prime factor affecting happiness. People are more likely to express a higher level of happiness when they live in walkable and pedestrian-friendly neighborhoods, especially those that are privileged with visual stimulation, aesthetic conditions, and perceived safety (Cao, 2016; Gim, 2020; Kent et al., 2017; Pfeiffer et al., 2020). This type of built environment provides residents with great opportunities to walk, meet, and socialize, which in turns strengthens the social relationship between neighbors, thus, making them happier (Diener & Seligman, 2002; Kent et al., 2017; Mason, 2010; Mouratidis, 2018).

On the other hand, living in a built environment with high automobile use rates leads to cognitive unhappiness due to exposure to air pollution and traffic congestion (Lenzen & Cummins, 2013). The frequent use of cars has been found to cause slight feelings of loneliness, while people who walk or cycle more often have greater access to a social relationship with neighbors and thus feel more social satisfaction (van de Coevering et al., 2016; Weijs-Perrée et al., 2015). Additionally, a regular car driving habit harms people's health and indirectly impacts their satisfaction and cognitive happiness (Frank et al., 2004; Gatersleben & Uzzell, 2007; Gottholmseder et al., 2009; Jacobson et al., 2011; Martin et al., 2014; Roberts et al., 2011). Regular driving results in a sedentary lifestyle that develops health problems such as obesity accompanied by several different chronic diseases like hypertension, cardiac disorders, vitamin deficiencies, and cancers that consequently cause lower levels of happiness and depression (Badawi & Farag, 2021; Bodea et al., 2009; Inyang & Stella, 2015; National Center on Health, 2016). The permanent exposure to air toxins due to excessive use of fossil fuel cars also causes various diseases, such as damage to neurological, reproductive, and immune systems (EPA, n.d.). Moreover, the increasing use of private cars causes an increase in traffic congestion in streets, which increases the mobility cost for people and the degradation of urban infrastructure. It negatively impacts environment, people health, and their comfort (Oleśków-Szłapka et al., 2020). Another supportive study found that pedestrians and cyclists also missed fewer workdays than their counterpart car drivers (Hansson et al., 2011).

Despite all the negative impacts of driving cars as an unsustainable mobility mode, there are positives. Cars may evoke a high positive feeling in some cases that should be recognized and considered by scholars and planners (Morris & Guerra, 2015). For example, a recent study lends support to the positive feeling of driving private cars, in which Aall et al. (2011) found that people prefer a private car over public transportation. Driving cars can be a cheaper mobility mode than public transportation, especially when driving to outdoor recreation destinations, and poor public transportation services disappoint travelers. Similarly, the limitation of effective mobility options and services can influence people's happiness by providing less access to life's necessities which may lead to disappointment and lower levels of happiness. For example, poor mobility options may reduce access to educational services, health-care facilities, employment opportunities, less interaction with the community, and isolation (Goodman, et al., 2021).

In sum, numerous studies support the notion of normalizing the sustainable mobility trajectory in people's travel behavior and attitude, which can contribute to higher levels of happiness. This emphasizes the link between happiness and sustainable mobility. However, the improper development of sustainable mobility modes can affect people's feelings, and therefore, their choices, making less sustainable modes more comfortable, affordable, safer, and preferable. Accordingly, the implication of sustainable mobility alternatives, plans, and measures should be effectively developed to be integrated into daily habits to contribute to affective and cognitive happiness while reinforcing the links with sustainability.

Branding cities through strengthening the link between sustainable mobility systems and cities' happy imagery could be a powerful strategy for promoting and developing sustainable mobility alternatives. International indexes are used to measure cities' happiness levels, contributing to branding the happiest cities to the global community. Below, I briefly provide an example of a city that uses happiness to define its sustainable mobility programs. Then, I review sustainable mobility's weighting factor within three international happiness indexes, which are the World Happiness Report, the Happy Planet Index, and the Better Life Index.

10.1.4 Branding Happy Cities Through Sustainable Mobility

The term *Branding City* refers to the process of communicating a city's image to the world by creating an emotional attachment to the high-quality urban features, strengthening the city's identity, and focusing on its comparative advantage. From an economic perspective, branding cities becomes crucial for cities to survive in a global perspective that has been imposed by economic competition (Winfield-Pfefferkorn, 2005). For example, emphasizing attractiveness and prosperity and a happy lifestyle attracted more tourists and boosted tourism-related business in Helsinki (Bei, 2014).

Branding a *Happy City* is a process of shaping an urban imaginary of a happy life in a certain city and a tool to make and market attractive places (Farag, 2019; Helmy, 2008). It can also be an opportunity to enhance the branded cities' role to go beyond the city's visual image to the multi-integrated associations between places, people, urban experiences, and sustainability.

Few political leaders have recognized the advantages of branding cities through strengthening the association between sustainable mobility systems and the city's happy imagery. One such leader is Peñalosa, the mayor of Bogotá, who sought public happiness by transforming into sustainable mobility. According to his vision, the city municipality issued several political reforms and established urban development projects to stop building highways. Instead, the municipality planned to rebuild Bogotá's streets and public spaces to be accessible for children and people who do not own private cars. The city invested more in installing high-quality sidewalks, pedestrian streets, parks, bicycle paths, and libraries. In addition, they developed a Bus Rapid Transit System to reduce car traffic and provide an affordable transport system for low-income people. The city's efforts intentionally aligned the sustainable mobility development with the term *happiness*, which seems more attractive language to the public. The applied strategies resulted in branding Bogotá as a happy city with a robust sustainable mobility system, making its experience a role model for other cities. Consequently, municipal officials from many countries visited Bogotá to closely investigate the applied measures that focused on happiness, sustainable mobility, and people. Other manifestations of success are the notable decrease in traffic accidents and crime and improved air quality in Bogotá's city (Wright & Montezuma, 2004).

10.1.5 Sustainable Mobility Metrics Within Happiness Assessment Indexes

The trend to brand a group of cities as the *World's Happiest* is growing, as they surpass others in happiness levels. In line with this trend, happiness standards become a top priority. The World Bank and United Nations have developed the happiness measurement progress by initiating a series of metrics to measure happiness levels. *The World Happiness Report*, the *Happy Planet Index*, and the *Better Life Index* are global initiatives that aim to measure cities' happiness levels (Farag et al., 2019). Additionally, many countries worldwide are setting national indexes for measuring and assessing happiness (Cloutier et al., 2014). Measuring and assessing happiness at the community level has become a branding strategy that combines subjective individual evaluation and objective metrics such as health, education, and economic prosperity (Haller & Hadler, 2006). Below, I review three international happiness indexes used to demonstrate the happiest cities worldwide, in which each index developed a set of metrics that differs in its priorities.

10.1.5.1 World Happiness Report (WHR)

The *World Happiness Report* is a broad survey that ranks 156 countries worldwide to assess how happy their people perceive themselves. The six variables included in the happiness assessment are 1. Levels of GDP, 2. Life expectancy, 3. Generosity, 4. Social support, 5. Freedom, and 6. Corruption, which have mainly focused on economic and socio-political aspects. Accordingly, the cities that score highest in those six variables are reported to have the happiest people worldwide. According to the WHR (2020) reports, the five happiest cities worldwide in 2020 are Helsinki, Finland (7.828); Aarhus, Denmark (7.625); Wellington, New Zealand (7.553); Zurich, Switzerland (7.541); and Copenhagen, Denmark (7.530), all of which are located in countries often ranked the happiest in the last several years (WHR, 2021). Sustainability and, more specifically, sustainable mobility, however, are not considered enough within these metrics.

The theme of the 2020 WHR was *environments for happiness*, in that its mission includes links between happiness and sustainable development. Yet, the six variables used to derive the Happiness Score do not include sustainable mobility or sustainability as a variable/metric. Chapter 6 in the WHR, 2020 report shows an interesting interpretation of how sustainable development is conducive to human happiness, emphasizing that those countries who are not doing well in terms of the Sustainable Development Goals (SDGs) likely have people who are experiencing more negative emotions (WHR, 2020). The void of sustainable mobility is interesting in that the United Nations General Assembly set SDG11.2 is aimed at sustainable mobility (SDG11.2).

10.1.5.2 Happy Planet Index (HPI)

The Happy Planet Index is one of the few International Happiness Indexes that measure happiness against sustainability. It combines four metrics which are 1. Experienced Wellbeing, 2. Life expectancy, 3. Inequality of outcomes, and 4. Ecological footprint. It aims to measure sustainable wellbeing and proposes that people can live good lives without overwhelming the Earth. Its metrics reflect the sustainability pillars so that the happiest countries in the HPI index are those where people live long and happy lives at little cost to the environment.

Some wealthy countries, such as the USA, do not score highly on the HPI overall, although they score high in wellbeing and life expectancy as the USA has one of the largest ecological footprints in the world. Conversely, Costa Rica ranks the happiest country worldwide, with a significantly smaller Ecological Footprint, following by Mexico, Colombia, Vanuatu, and Vietnam, respectively (Happy Planet Index, 2016). Switzerland and Finland, two of the top Happiest countries in (WHR, 2021), ranked the 24th and 37th in the HPI, respectively. However, they are among the highest list scored in the metrics of Experienced Wellbeing and Inequality of Outcomes, and it is the Ecological Footprint factor that probably impacts the total score most (Happy

Planet Index, 2016). Although this index put considerable weight on sustainability, it still lacks a sustainable mobility indicator.

10.1.5.3 Better Life Index (BLI)

The Better Life Index is designed to involve people in measuring countries' wellbeing and empowering them to decide which metric matters more in the assessment process. Eleven metrics are used to assess countries' happiness levels, and each person can frame the assessment system by providing an individual weight for each metric fitting their preference. The metrics used to make a better life for people are 1. Housing, 2. Income, 3. Jobs, 4. Community, 5. Education, 6. Environment, 7. Civic engagement, 8. Health, 9. Life satisfaction, 10. Safety, and 11. Work-life Balance. Each metric is based on one to three equal weight indicators. A countries' rank is subjectively determined by combining the weights for 11 metrics, individually. The BLI contains both subjective and objective assessments for countries' various life aspects; however, the index is limited to Organisation for Economic Co-operation and Development (OECD) countries. Of note, the 11 metrics used to assess the happiness levels do not include sustainable mobility explicitly, but the tenth metric, *Safety*, includes a minor indicator for (Feeling safe walking alone at night) which is a primary attribute of sustainable mobility (OECD Better Life Index, 2021).

In sum, the role of international happiness indexes in branding cities has widely been recognized in recent years, affecting several aspects of cities' lives. The branding process could be an effective tool to emphasize a link between happiness and sustainable mobility and maximize the branded cities' role in improving people's lives and saving the environment, making people happier. The three international indexes (WHR, HPI, and BLI) aimed to assess happiness in cities and countries depending on various metrics and indicators such as housing affordability, ecological footprint, freedom, and safety. However, none of them have explicitly and decisively considered sustainable mobility as a possible indicator of communities' happiness. Therefore, this chapter assumes that investigating the sustainable mobility practices in the cities branded the happiest can reveal reflective insights of links between sustainable mobility and happiness, also providing future direction for such measures.

10.2 Methods and Procedures

This chapter aims to link sustainable mobility to the happiness levels of urban people. The literature review confirms the key role of sustainable mobility contribution to making people happier, and discusses the power of branding cities by strengthening the link between sustainable mobility systems and cities' happy imagery on promoting sustainable mobility measures. However, the international happiness indexes, which are relied on to measure and demonstrate the world's happiest cities, have not yet considered sustainable mobility as a possible indicator of communities' happiness. Therefore, a mixed methodology is used to investigate the link in the cities officially branded as the happiest in the world, drawing on quantitative and qualitative approaches. The effort confirms that those cities, who endeavor to achieve higher levels of happiness must enhance the quality of life of people through developing sustainable mobility measures.

10.2.1 Quantitative Methods

Within the quantitative approach, I use a descriptive statistics method for data analysis, aiming to investigate the Sustainable Mobility Performance of ten cities branded the happiest, as follows:

a. Ten cities are selected, which are located in countries that are both consistently ranked among the top 10 happiest in the last five years (WHR, 2020, 2021). The data availability, continuity of deployment, and wide prevalence of the WHR index make it a proper reference for selecting the purposive samples of the happiest cities.
b. The list of ten cities is shown in Table 10.1, below, is arranged in descending order according to the happiness score (WHR, 2020) so that the first city on the list is the highest in happiness levels, and so on.
c. The sustainable mobility performance of each city is demonstrated by reference to five International Indexes allotted in Sustainable Mobility. The fives indexes are:

    i. Sustainable Cities Mobility Index (Arcadis, 2017)
    ii. Transport Global Tracking Framework 2.0 (Sum4all, n.d.)
    iii. Urban Mobility Index (Cebr, 2017)
    iv. Urban Mobility Index (Here) (Urban Mobility Index.here, 2018)
    v. Arthur D. Little Urban Mobility Index 3.0 (Arthur D. Little, 2018)

    All five indexes provide statistical data of global cities' performance in sustainable mobility in the form of scores and ranks. All available data of the ten selected cities is illustrated in Table 10.1; some indexes have not evaluated some cities; therefore, a dash is used when the city evaluation data is not available.

d. The five indexes evaluate the global cities' performance by ranking them. Those ranks are used in Table 10.1 to show the happiest cities' progress in sustainable mobility globally.
e. Only four indexes provide a score, in the form of a percentage (%), to evaluate each city's sustainable mobility performance. Urban Mobility Index-Here does not provide scores, and therefore, this index score's column is removed.
f. An additional column is added to Table 10.1 to find the average of each city's performance according to the available scores. For each city's performance average, 2–4 scores from different indexes are used, according to the data availability.

Table 10.1 Sustainable mobility performance in cities branded the happiest

Happy cities	Sustainable index									
	Sustainable cities mobility Index		Global tracking framework/Country level		Urban mobility index		Urban Mobility index (Here)	Arthur D. Little urban mobility index 3.0		Index score-average %
	Index score %	Rank n = 100	Index score %	Rank n = 183	Index score %	Rank n = 35	Rank n = 38	Index score %	Rank n = 100	
Helsinki, Finland (7.828)	–	–	75.6	14	–	–	9	52.8	10	64.2
Zurich, Switzerland (7.541)	65	2	76.8	11	65	6	5	53.2	9	65
Copenhagen, Denmark (7.530)	61	12	77.6	10	69	4	1	54.6	4	65.55
Oslo, Norway (7.464)	–	–	77.9	9	80	1	–	–	–	78.95
Stockholm, Sweden (7.373)	62.7	9	84.9	3	–	–	–	–	–	73.8
Brisbane, Australia (7.337)	47.7	47	61.1	35	–	–	–	–	–	54.4
Toronto Metro, Canada (7.298)	46.1	54	57.6	44	51	15	20	44.7	31	49.85

(continued)

Table 10.1 (continued)

Happy cities	Sustainable index									
	Sustainable cities mobility Index		Global tracking framework/Country level		Urban mobility index		Urban Mobility index (Here)	Arthur D. Little urban mobility index 3.0		Index score-average %
	Index score %	Rank n = 100	Index score %	Rank n = 183	Index score %	Rank n = 35	Rank n = 38	Index score %	Rank n = 100	
Melbourne, Australia (7.296)	45.9	55	61.1	35	43	26	29	43.2	40	48.3
Perth, Australia (7.253)	37	87	61.1	35	–	–	–	–	–	49.05
Sydney, Australia (7.133)	47.3	51	61.1	35	–	–	30	42.2	48	50.2

g. Two illustrations are added for data analysis; Chart 10.1 and Map 10.1. Chart 10.1 is used to show the correlation between the average of the sustainable mobility performance of each city to the city's happiness levels. The link between sustainable mobility and happiness levels is emphasized.

h. Map 10.1 shows the ten happiest cities allocated and coded according to their sustainable mobility performance. This demonstrates the geographical distribution of the happiest cities that explicitly and decisively confirm the link and could

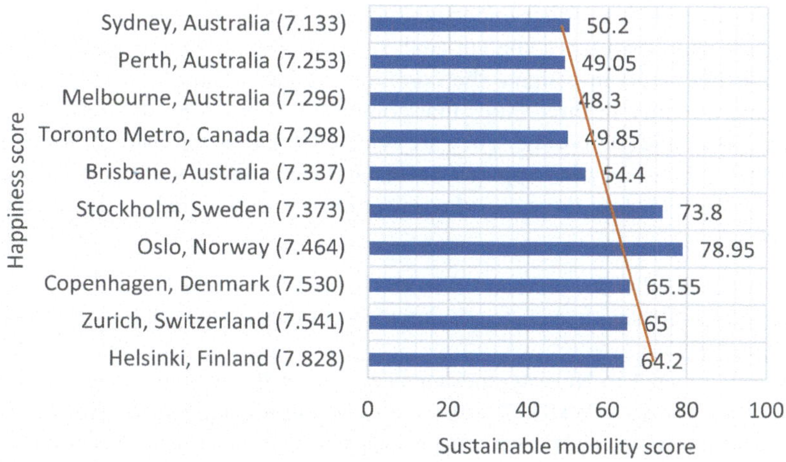

Chart 10.1 Sustainable mobility performance (Average) in 10 cities branded the happiest

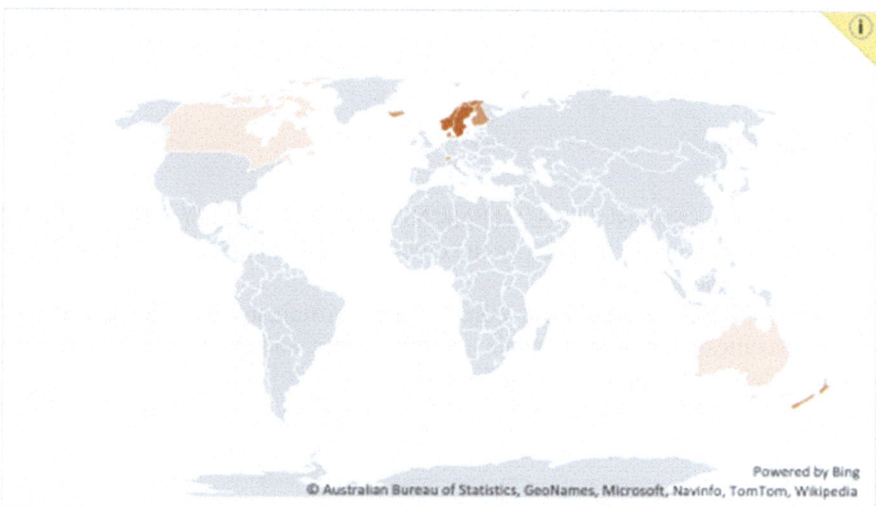

Map 10.1 Cities branded the happiest coded bsy sustainable mobility performance. *Recourse* The researcher, 2021

be a promising start of branding those cities' happy imagery within sustainable mobility performance and practices.

10.2.2 *Qualitative Methods*

The qualitative approach aims to correlate the sustainable mobility measures and the level of happiness in three cities officially branded the happiest. The qualitative approach provides further understanding and validation of the results from the quantitative analysis. I use a content analysis method with the statistical and documentary materials to investigate the common practices and measures to promote sustainable mobility in three cities that contribute to branding them the happiest: Helsinki, Oslo, and Melbourne. They are selected from the ten listed cities in Table 10.1 as those cities vary in happiness levels and sustainable mobility performance. Finally, as shared below, there are other intriguing reasons to include the three in my analysis:

1. *Helsinki* is officially announced as the happiest city in 2020 and is located in Finland, which has been listed the happiest country for the last four years (WHR, 2020, 2021).
2. *Oslo* is located in Norway and has been one of the happiest cities worldwide for several years (WHR, 2020). Due to its significant efforts with respect to sustainable measures, Oslo has been awarded the title of European Green Capital 2019 (European Union, 2018).
3. *Melbourne* in Australia represents one of the top listed cities in happiness worldwide (WHR, 2020). Additionally, it has been awarded as the *World's Most Liveable City* for seven years in a row as it offers the best quality of life for its people (EIU, 2021).

10.3 Results

I have arranged the results in three sections below, given the explored links between happiness and sustainable mobility in the cities officially branded as the happiest in the world. The findings are:

(1) the happiest cities ranked and scored higher in sustainable mobility indexes, (2) the happiest cities promote sustainable mobility modes, and (3) sustainable mobility links to the levels of happiness in the happiest cities.

10.3.1 The Happiest Cities Ranked and Scored Higher in Sustainable Mobility Indexes

According to Table 10.1, happy cities obtain high scores and ranks in sustainable mobility indexes. Specifically, those cities with the highest happiness scores in the WHR index are ranked the highest among other cities assessed by the five indexes. For example, Oslo is ranked first in the Urban Mobility Index, followed by Copenhagen, the fourth, among 35 cities. Copenhagen is ranked first in the Urban Mobility Index-Here, followed by Zurich, the fifth among 38 cities. Zurich is ranked second in the Sustainable Cities Mobility Index, followed by Stockholm, the ninth among 100 cities. Furthermore, the cities rank high places among 100 cities in the Arthur D. Little Urban Mobility Index 3.0, in which Copenhagen ranked fourth, followed by Zurich, the ninth. Similarly, most cities are ranked high in Global Tracking Framework in a comparison between 183 countries, while Stockholm ranked third, followed by Oslo, the ninth.

Additionally, Oslo and Stockholm cities show a statistical significance in the average of sustainable mobility scores. In contrast, Australia and Canada's cities, which come later in the Happiest cities list, have lower average values of sustainable mobility scores. However, Helsinki, the happiest city worldwide, shows a moderate significance of the sustainable mobility average, which needs further investigation. In sum, some happiest cities' performance is notable in sustainable mobility, such as Oslo, while others are still on the right track, such as Melbourne.

Even though the happiest cities received varied evaluations and consequently various ranks and scores due to the variation of each index's metrics, the consistent repetition of high ranks with slight differences confirms the prominent performance of happy cities in sustainable mobility.

Chart 10.1 displays the sustainable mobility average for ten cities branded the happiest worldwide, showing a linear correlation between the happiness score and sustainable mobility scores average, with slight deviation, represented in Oslo and Stockholm cities' averages. Oslo has demonstrated advanced progress in sustainable mobility ahead of Helsinki, the first in happiness. However, the first five cities with the highest happiness levels score higher in sustainable mobility than other cities. The finding requires further investigation, in which the happy cities' performances and practices in sustainable mobility are compared to those cities' performance with the lowest happiness levels.

Map 10.1 shows the ten happiest cities coded according to their sustainable mobility performance. The happiest cities with notable performance in sustainable mobility significantly occupy limited areas, assuring that there is a great effort needed to foster happsstainable mobility across the globe.

Accordingly, the data analysis shows an explicit link between happiness levels and sustainable mobility in that the cities branded the happiest worldwide reported higher ranks and scores in sustainable mobility indexes. This finding is consistent with the reviewed studies, which concluded that cities with higher reported scores in sustainability indices could offer their residents more significant opportunities for happiness (Cloutier & Pfeiffer, 2015; Leyden et al., 2011; Pfeiffer & Cloutier, 2016; Veenhoven, 1993; White et al., 2013, 2017; Zidansek, 2007). It is also consistent with a recent study, which concluded that American cities with higher use of sustainable mobility modes and a dedicated commitment to improving sustainable mobility infrastructure have higher happiness levels among residents (Cloutier et al., 2017). The finding contributes to research, providing insights into sustainable mobility's potentials to make people happier. Below, I investigated the common measures to promote sustainable mobility modes in three cities branded the happiest, revealing thoughtful insights of links between sustainable mobility and happiness.

10.3.2 The Happiest Cities Promote Sustainable Mobility Modes

One of the common measures in the three cities is that they have made serious and long-term plans and actions to promote sustainable mobility modes that, according to the literature review, could contribute to affective and cognitive happiness while reinforcing links with sustainability.

Helsinki started a gradually developing process after the Second World War, including all living standards (Weckroth, 2012), resulting in a plan toward sustainable mobility modes. It has implemented several strategies, especially ones concerning promoting cycling to transform into a cycle-friendly city where the mode constitutes 10% of all trips and 70% of the city's citizens cycle at least sporadically. In order to reach these levels, the city affords several facilities and services to improve the built environment and infrastructure for cyclists, such as extending the cycling network, improving bicycle parking facilities, and bicycle self-repair facilities. However, it aspires to reach Copenhagen's level, one of the happiest cities and the leading cycling city in Europe, where 30% of all trips are made by bicycle. Regarding other mobility modes, Helsinki daily trips reflect significant progress of transforming into sustainable mobility where 37% of all trips are made by walking and 30% by public transport, while private cars constitute only 22% of all trips. As part of sustainable mobility promotion, Helsinki issued a recent policy to integrate different sustainable modes; it allows cyclists to take their bike when they ride the metro or train at all times of the day and supports this initiative by improving the parking facilities near public transport stations. In addition, Helsinki establishing new facilities for vehicle sharing systems (Helsinki City Planning Department, 2017).

Similarly, *Oslo's* experience in sustainable mobility is long and has been gradually implemented since the early 1970s (Doheim et al., 2020b). Oslo aspires to be a leading sustainable city in the world. To achieve its endeavor, it has adopted sustainable mobility as one of the key directions of sustainability, as it has achieved many changes in urban mobility patterns in the last several years. Several strategies are applied to improve walking and cycling conditions and enhance public transport to be fast, clean, and accessible. Accordingly, Oslo provides its citizens with the longest network of bicycle tracks, designed with green corridors, to make cycling an attractive mobility mode (Doheim et al., 2020a, 2020b; Oleśków-Szłapka et al., 2020).

Additionally, *Oslo* provides innovative ways to promote sustainable mobility. One of its emerging projects in 2017 is the *bicycle hotel*, designed to be a luxury welcoming environment for cyclists providing closed safe parking space for bicycles that one gains entry through a smartphone application. More, Oslo develops new smart and sustainable mobility modes such as electric kick scooters and leads other European cities to apply clean electric vehicles, including cars and bikes, to be affordable to all citizens and afford them financial incentives to buy and use them. In 2018, over 60% of cars sold in the Oslo region were Electrical Vehicles (European Union, 2018). Oslo has taken several actions to promote sustainable mobility while enhancing people's quality of life. For example, it replaced over 700 parking spots and 35 miles of roads with bike tracks and green spaces to free streets and open spaces for people and make them safer for pedestrians and cyclists. It runs the local bike-share project Urban Sharing to cope with the new needs and enhance the vehicle sharing experience in Oslo (Doheim et al., 2020a, 2020b). Consequently, the sustainable mobility modes become relatively more competitive than the private car. As a result, a notable increase of 50% of travel trips in trams, train, and metro, and public transport journeys is documented, since 2007. And many young people use car-sharing, scooters, and urban bikes as an everyday mobility mode (Oleśków-Szłapka et al., 2020). The significant efforts made by Oslo city to enhance the travelers' experience with sustainable mobility support its high score of both sustainable mobility and happiness.

On the other hand, *Melbourne* is on the right track to promote sustainable mobility despite several challenges. Melbourne is predominantly low-density suburban in nature, radiating outwards from the original Victorianera inner-city core (Goodman et al., 2021). The prevailing mobility in old Melbourne in 1947 was public transport, including electric trams and trains that were used by 80% of urban travelers in the inner areas of the city, while many suburban trips were local and made on foot or bicycle (Moriarty & Honnery, 2008).

Currently, the urban form of Melbourne affects the development of mobility patterns, causing several challenges. The main challenges are urban sprawling and low urban density in Melbourne's outer suburbs, which leads to high dependence on private cars. The car weight in Melbourne's city is increased compared to other modes of mobility to represents 40% of all trips (Fernandez-Sanchez & Fernandez-Heredia, 2018). This travel attitude, combined with the rapid growth of population, has caused pressures on the city's infrastructure (Horan et al., 2014). The urban infrastructure and mobility services have not kept pace with the rapid population growth in many areas, causing high levels of traffic congestion, making people waste an unreasonable amount of time daily (Goodman et al., 2021).

Moreover, the transport sector is considered the second prominent reason for the Greenhouse gas emissions in Melbourne. In fact, congestion in Melbourne and some other Australian cities cost Australia from $12.8 Billion in 2010 to $16.5 Billion in 2015, with projected increases to about $30 Billion in 2030. The increased traffic congestion and decreased environmental quality, which are both main factors in assessing the city's livability, would undermine people's happiness and wellbeing (Horan et al., 2014).

However, Melbourne works on implications for policy development and urban design interventions to encourage greater use of sustainable modes of mobility (Taylor & Thompson, 2019). It implemented several policies and measures to improve cycling networks and services, promote cycling and make it more accessible for all people. Currently, 51 bicycle stations are primarily located in the inner-city of Melbourne, in addition to over 600 bicycles operating in the inner city. The aforementioned challenges that hinder sustainable mobility development in Melbourne may predict its average scores in sustainable mobility indexes. Therefore, a more sustainable alternative is required to overcome these challenges and their possible future consequences.

10.3.3 Sustainable Mobility Links to the Levels of Happiness in the Happiest Cities

The three cities in Sect. 10.3.2 promote sustainable mobility modes and have made several actions and issued policies, pursuing higher levels of happiness among citizens.

Helsinki has a long-term and comprehensive strategic plan for sustainable mobility that considers the mobility system as a whole, including all mobility modes. Its efforts result in high levels of satisfaction among citizens where 96% are happy with the measures taken to promote cycling, while 80% are satisfied with public transport. However, further progress is needed to cope with the contemporary challenges, such as the increasing rate of commuting trips to Helsinki where cars drive from other municipalities, which leads to a crucial need for developing plans at the regional level (Willberg, 2019). However, Helsinki, the happiest city worldwide, still represents balanced progress and conscious efforts in investment in sustainable mobility infrastructure and services, supporting the notion of a link between happiness and sustainable mobility.

Oslo's experience is evidence of a sustainable city fostering happiness through developing sustainable mobility, as the taken action works for the city's people's happiness as a priority. Besides its endeavor and significant efforts, Oslo demonstrated its banner to be a "City for Everyone, Putting People First." Oslo is one of the few cities that has not witnessed public resistance to the measures taken to promote sustainable mobility due to its commitment to take into account its community's satisfaction and involving people in the process of decision making from the beginning (Doheim et al.,). The notion of improving mobility with people and for people enhances Oslo city's environment, attractiveness, and quality of life, which in turn improves people's satisfaction and happiness levels. This results in high scores of sustainable mobility and happiness.

In Melbourne, most people prefer private cars over public transport due to the lack of public mobility services and the absence or poor quality of public transport in Melbourne's outer growth areas, and the poor walkability for public transport. High levels of walkability are only present in the inner suburbs of Melbourne. Due to the centralized job distribution in Melbourne, people who live in the outer areas of the city spend not less than two hours a day commuting to work, consuming the time they can spend with their families and friends. The lack of sustainable mobility services influences several life choices, including where people work, study, their involvement in the community, their physical and mental health, and their relationships with friends and family. In these areas, with poor public transport and walkability, private cars' use becomes necessary, increasing sedentary lifestyle and unsustainable mobility modes associated with lower levels of happiness than those using sustainable mobility modes (Goodman et al., 2021).

Melbourne comes later in the happiest list after Heliniski and Oslo, as well as its score in Sustainable Mobility Performance average, which is lesser than both. Melbourne still struggles to apply sustainable mobility measures, depending mainly on private cars. This finding supports the notion that cities, who implement sustainable mobility measures invest more in their people's comfort, wellbeing, and happiness.

Finally, the results support the literature review in the key role sustainable mobility plays in making people happier. Additionally, the study confirms the power of branding the happiest cities, also using sustainable mobility across the globe. In short, the happiest cities invest in their people's happiness by focusing on sustainable mobility. The study is limited, however, in that the happiest cities selected for this study and their evaluations are based on statistical data from five indexes. More, some of the happiest cities (Aarhus, Denmark; Wellington, New Zealand; Bergen, Norway; and Reykjavik, Iceland) are excluded from this study due to the unavailability of and limited access to their data—further data generation and research is needed.

The chapter raised several concerns as potential areas for research. First, the high performance of the happiest cities also requires a focus on those cities with the lowest happiness levels and their performance in terms of sustainable mobility. A comparative study between two samples of cities representing the highest and lowest happiness levels might clearly emphasize the link between happiness and sustainable mobility. Additionally, branding happy cities is a powerful strategy to emphasize sustainable practices globally. An approach combining cities with higher levels of happiness and higher performance of sustainable mobility could also offer research opportunities to investigate future branding tools, methods, and practices.

Finally, the COVID-19 pandemic has impacted the quality of life and public health worldwide, causing many challenges to cities' happiness and sustainable mobility practices. Future challenges for promoting sustainable mobility may include the avoidance of public transport in favor of social distance. The new challenges orient researchers in the field to investigate new solutions enhancing happiness and public health, while also promoting sustainable mobility modes.

10.4 Conclusion

This chapter contributes to research indicating that sustainable mobility provides significant potential to make people happier. The literature review demonstrates that sustainable mobility impacts people's quality of life, public health, and their happiness. It also emphasizes the power of branding cities by strengthening the link between sustainable mobility systems and cities' happy imagery through the promotion of sustainable mobility measures. A review of the international happiness indexes that assess and indicate the world's happiest cities revealed that sustainable mobility was not considered as a possible indicator of communities' happiness. Therefore, a mixed methodology was used to investigate the link between the cities officially branded as the happiest in the world and sustainable mobility. The results indicated that those cities, who hope to achieve higher levels of happiness must enhance the quality of life by developing sustainable mobility measures. More, the improper development of sustainable mobility facilities and infrastructure can affect people's affective and cognitive happiness, making their choices biased toward less sustainable modes such as private cars. Accordingly, the implication of sustainable mobility alternatives, plans, and measures should be integrated into daily habits to contribute to affective and cognitive happiness while reinforcing links with sustainability.

Even though the included happiness indexes (WHR, HPI, and BLI) do not consider sustainable mobility in their metrics, the link between happiness and sustainable mobility is noted in those cities branded the happiest. It might be that bringing happiness to people in cities through social, economic, and political measures never contradicts the provision of a sustainable mobility system that is safe, affordable, clean, and accessible. On the contrary, happiness-enhancing efforts must integrate mobility quality supporting daily life necessities. Therefore, emphasizing sustainable mobility in happiness index reports is crucial to encouraging more sustainable mobility measures globally.

Finally, combining sustainable mobility and happiness-enhancing efforts might slightly change the world happiness map, moving some cities and countries to the top of the list and some sedentary lifestyle cities and countries to the bottom. However, most important is that this trend could potentially encourage cities to aspire toward higher levels of happiness by promoting sustainable mobility, thus saving the environment and enhancing people's lifestyles, health and happiness. Ultimately, happiness is a possible element that can be more deeply embedded into the discourse of promoting sustainable mobility. I recommend considering sustainable mobility in future happiness metrics and indexes, which in turn could place a sustainable and happy future in cities' development plans and actions.

References

Aall, C., Klepp, I. G., Engeset, A. B., Skuland, S. E., & Støa, E. (2011). Leisure and sustainable development in Norway: Part of the solution and the problem. *Leisure Studies, 30*(4), 453–476.

Arcadis. (2017). *Sustainable cities mobility index 2017: Bold moves*. https://trid.trb.org/view/1489272.

Argyle, M. (1997). Is happiness a cause of health? *Psychology and Health, 12*(6), 769–781.

Badawi, S., & Farag, A. A. (2021). Young Saudi women's travel behavior change over 2015/2020. *Journal of Transport & Health, 21*, 101080.

Bei, J. (2014). *Place marketing: Cases Shanghai and Helsinki*. [Bachelor's thesis, Kymenlaakson ammattikorkeakoulu University of Applied Sciences]. https://www.theseus.fi/handle/10024/75467.

Berejnoi, E., Ulluwishewa, R., Cloutier, S., Gibbons, L., Puga, S., & Uthumange, A. (2019). Spirituality: The missing link of sustainability and happiness as a framework for holistic development. In Y. Kee, S. J. Lee, & R. Phillips (Eds.), *Perspectives on community Well-Being* (pp. 203–224). Springer. https://doi.org/10.1007/978-3-030-15115-7_11.

Black, S. V., Cooper, R., Martin, K. R., Brage, S., Kuh, D., & Stafford, M. (2015). Physical activity and mental wellbeing in a cohort aged 60–64 years. *American Journal of Preventive Medicine, 49*(2), 172–180.

Bodea, T. D., Garrow, L. A., Meyer, M. D., & Ross, C. L. (2009). Socio-demographic and built environment influences on the odds of being overweight or obese: The Atlanta experience. *Transportation Research Part a: Policy and Practice, 43*(4), 430–444.

Cao, X. J. (2016). How does neighborhood design affect life satisfaction? Evidence from Twin Cities. *Travel Behaviour and Society, 5*, 68–76.

Cebr. (2017). *Urban Mobility Index*. https://www.qualcomm.com/media/documents/files/urban-mobility-index-report.pdf.

Cloutier, S., Karner, A., Breetz, H. L., Toufani, P., Onat, N., Patel, S., Paralkar, S., Berejnoi, E., Morrison, B. A., Papenfuss, J. & Carlson, C. (2017). Measures of a sustainable commute as a predictor of happiness. *Sustainability, 9*(7), 1214

Cloutier, S., & Pfeiffer, D. (2015). Sustainability through happiness: A framework for sustainable development. *Sustainable Development, 23*(5), 317–327.

Cloutier, S., Larson, L., & Jambeck, J. (2014). Are sustainable cities "happy" cities? Associations between sustainable development and human wellbeing in urban areas of the United States. *Environment, Development and Sustainability, 16*(3), 633–647.

Deeg, D. J., & van Zonneveld, R. J. (1989). Does happiness lengthen life? The prediction of longevity in the elderly. In R. Veenhoven (Ed.), *How harmful is happiness: Consequences of enjoying life or not* (pp. 29–43). https://personal.eur.nl/veenhoven/Pub1980s/89a-con.htm.

Diener, E., & Seligman, M. E. (2002). Very happy people. *Psychological Science, 13*(1), 81–84.

Doheim, R. M., Farag, A. A., & Kamel, E. (Eds.). (2020). *Humanizing cities through car-free city development and transformation*. IGI Global.

Doheim, R. M., Farag, A. A., & Badawi, S. (2020). Success measures for transforming into car-free cities: Recommendations for implementation. In R. M. Doheim, A.A. Farag, & E. Kamel, (Eds.), *Humanizing cities through car-free city development and transformation* (pp. 232–268).

Duarte, A., Garcia, C., Giannarakis, G., Limao, S., Polydoropoulou, A., Litinas, N., (2010). New approaches in transportation planning: Happiness and transport economics. *NETNOMICS: Economic Research and Electronic Networking, 11*(1), 5–32.

EIU. (2021). *The Global liveability index*. https://www.eiu.com/topic/liveability.

EPA. (n.d.). *Research on health effects, exposure, and risk from mobile source pollution*. https://www.epa.gov/mobile-source-pollution/research-health-effects-exposure-risk-mobile-source-pollution.

European Union. (2018). *OSLO: European Green Capital 2019*. Bietlot in Belgium. https://ec.europa.eu/environment/europeangreencapital/wp-content/uploads/2019/Oslo%20Brochure_EGCA%202019.pdf.

Farag, A. A. (2019). The story of NEOM city: Opportunities and challenges. In S. Attia, Z. Shafik, & A. Ibrahim (Eds.), *New cities and community extensions in Egypt and the Middle East* (pp. 35–49). Springer International Publishing. https://link.springer.com/chapter/https://doi.org/10.1007/978-3-319-77875-4_3.

Farag, A. A., Badawi, S. R., & Doheim, R. M. (2019). Assessment of user happiness in campus open spaces. *The Journal of Public Space, 4*(1), 45–64.

Fernandez-Sanchez, G., & Fernandez-Heredia, A. (2018). Strategic thinking for sustainability: A review of 10 strategies for sustainable mobility by bus for cities. *Sustainability, 10*(11), 4282.

Forsyth, A. (2015). What is a walkable place? The walkability debate in urban design. *Urban Design International, 20*(4), 274–292.

Frank, L. D., Andresen, M. A., & Schmid, T. L. (2004). Obesity relationships with community design, physical activity, and time spent in cars. *American Journal of Preventive Medicine, 27*(2), 87–96.

Gallo, M., & Marinelli, M. (2020). Sustainable mobility: A review of possible actions and policies. *Sustainability, 12*(18), 7499.

Gatersleben, B., & Uzzell, D. (2007). Affective appraisals of the daily commute: Comparing perceptions of drivers, cyclists, walkers, and users of public transport. *Environment and Behavior, 39*(3), 416–431.

Gim, T. H. T. (2020). The relationship between overall happiness and perceived transportation services relative to other individual and environmental variables. *Growth and Change, 51*(2), 712–733.

Goodman, R., Kroen, A., & Davern, M. (2021). Quality of life, sustainability and transport: The case of Melbourne, Australia. In J. Martinez, C. A. Mikkelsen, & R. Phillips (Eds.), *Handbook of quality of life and sustainability* (pp. 203–226). Springer.

Gottholmseder, G., Nowotny, K., Pruckner, G. J., & Theurl, E. (2009). Stress perception and commuting. *Health Economics, 18*(5), 559–576.

Guell, C., & Ogilvie, D. (2015). Picturing commuting: Photovoice and seeking wellbeing in everyday travel. *Qualitative Research, 15*(2), 201–218.

Haller, M., & Hadler, M. (2006). How social relations and structures can produce happiness and unhappiness: An international comparative analysis. *Social Indicators Research, 75*, 169–216.

Hansson, E., Mattisson, K., Björk, J., Östergren, P. O., & Jakobsson, K. (2011). Relationship between commuting and health outcomes in a cross-sectional population survey in southern Sweden. *BMC Public Health, 11*(1), 1–14.

Happy Planet Index. (2016). About the HPI. http://happyplanetindex.org/about.

Helsinki City Planning Department (2017). Pyöräilykatsaus 2017, Helsinki suunnittelee, 3.

Helmy, M. (2008). *Urban branding strategies and the emerging Arab cityscape: The Image of Gulf City*. University of Stuttgart.

Hoen, P. W., Denollet, J., de Jonge, P., & Whooley, M. A. (2013). Positive affect and survival in patients with stable coronary heart disease: Findings from the Heart and Soul Study. *The Journal of Clinical Psychiatry, 74*(7), 716–722.

Holden, E., Gilpin, G., & Banister, D. (2019). Sustainable mobility at thirty.*Sustainability, 11*(7), 1965.

Horan, E., Craven, J., & Goulding, R. (2014). Sustainable urban development and liveability. How can Melbourne retain its title as the world's most liveable city and strive for sustainability at the same time? *European Journal of Sustainable Development*, *3*(4), 61–61.

Inyang, D. M. P., & Stella, O. O. (2015). Sedentary lifestyle: Health implications. *IOSR Journal of Nursing and Health Science, 4*(2), 20–25.

Jacobson, S. H., King, D. M., & Yuan, R. (2011). A note on the relationship between obesity and driving. *Transport Policy, 18*(5), 772–776.

Kent, J. L., Ma, L., & Mulley, C. (2017). The objective and perceived built environment: What matters for happiness? *Cities & Health, 1*(1), 59–71.

Kim, D. (2018). The effects of urbanisation level and mobility on happiness in OECD countries. *International Journal of Happiness and Development, 4*(3), 195–208.

Lee, H. W., & Shin, J. T. (2020). The Effect of preferred physical activity on emotion, stress and happiness of university student. *Research in Dance and Physical Education, 4*(3), 63–70.

Lenzen, M., & Cummins, R. A. (2013). Happiness versus the environment—A case study of Australian lifestyles. *Challenges, 4*(1), 56–74.

Leyden, M., Goldberg, A., & Michelback, P. (2011). Understanding the pursuit of happiness in ten major cities. *Urban Affairs Review, 47*, 861–888.
Little, A. D. (2018). *The Future of mobility 3.0.* https://www.adlittle.com/futuremobilitylab/assets/file/180330_Arthur_D.Little_&_UITP_Future_of_Mobility_3_study.pdf.
Mason, S. G. (2010). Can community design build trust? A comparative study of design factors in Boise. *Idaho Neighborhoods. Cities, 27*(6), 456–465.
Martin, A., Goryakin, Y., & Suhrcke, M. (2014). Does active commuting improve psychological wellbeing? Longitudinal evidence from eighteen waves of the British Household Panel Survey. *Preventive Medicine, 69*, 296–303.
Morris, E. A. (2015). Should we all just stay home? Travel, out-of-home activities, and life satisfaction. *Transportation Research Part a: Policy and Practice, 78*, 519–536.
Morris, E. A., & Guerra, E. (2015). Mood and mode: Does how we travel affect how we feel? *Transportation, 42*(1), 25–43.
Moriarty, P., & Honnery, D. (2008). Low-mobility: The future of transport. *Futures, 40*(10), 865–872.
Mouratidis, K. (2018). Built environment and social wellbeing: How does urban form affect social life and personal relationships? *Cities, 74*, 7–20.
Mytton, O. T., Panter, J., & Ogilvie, D. (2016). Longitudinal associations of active commuting with wellbeing and sickness absence. *Preventive Medicine, 84*, 19–26.
National Center on Health. (2016). Sedentary lifestyle is dangerous to your health. NCHPAD. http://www.nchpad.org/403/2216/Sedentary~Lifestyle~is~Dangerous~to~Your~Health.
OECD Better Life Index. (2021). Accessed online on March 14, 2021. In http://www.oecdbetterlifeindex.org/.
O'Brien, C. (2008). Sustainable happiness: How happiness studies can contribute to a more sustainable future. *Canadian Psychology/psychologie Canadienne, 49*(4), 289–295. https://doi.org/10.1037/a0013235
Oleśków-Szłapka, J., Pawłyszyn, I., & Przybylska, J. (2020). Sustainable urban mobility in Poznan and Oslo—Actual state and development perspectives. *Sustainability, 12*(16), 6510.
Olsson, L. E., Gärling, T., Ettema, D., Friman, M., & Fujii, S. (2013). Happiness and satisfaction with work commute. *Social Indicators Research, 111*(1), 255–263.
Pfeiffer, D., Ehlenz, M. M., Andrade, R., Cloutier, S., & Larson, K. L. (2020). Do neighborhood walkability, transit, and parks relate to residents' life satisfaction? Insights from Phoenix. *Journal of the American Planning Association, 86*(2), 171–187.
Pfeiffer, D., & Cloutier, S. (2016). Planning for happy neighborhoods. *Journal of the American Planning Association, 82*(3), 267–279.
Ramanathan, S., O'Brien, C., Faulkner, G., & Stone, M. (2014). Happiness in motion: Emotions, wellbeing, and active school travel. *Journal of School Health, 84*(8), 516–523.
Rasciute, S., & Downward, P. (2010). Health or happiness? What is the impact of physical activity on the individual? *Kyklos, 63*(2), 256–270.
Richards, J., Jiang, X., Kelly, P., Chau, J., Bauman, A., & Ding, D. (2015). Don't worry, be happy: Cross-sectional associations between physical activity and happiness in 15 European countries. *BMC Public Health, 15*(1), 1–8.
Roberts, J., Hodgson, R., & Dolan, P. (2011). "It's driving her mad": Gender differences in the effects of commuting on psychological health. *Journal of Health Economics, 30*(5), 1064–1076.
Ruiz-Ariza, A., de la Torre-Cruz, M. J., Redecillas-Peiro, M. T., & Martinez-Lopez, E. J. (2015). Influence of active commuting on happiness, wellbeing, psychological distress and body shape in adolescents. *Gaceta Sanitaria, 29*(6), 454–457.
Schimmel, J. (2009). Development as happiness: The subjective perception of happiness and UNDP's analysis of poverty, wealth and development. *Journal of Happiness Studies, 10*, 93–111.
SDG11.2, Goal 11: Make cities inclusive, safe, resilient and sustainable. *Sustainable Development Goals.* https://www.un.org/sustainabledevelopment/cities/
Steptoe, A. (2019). Happiness and health. *Annual Review of Public Health, 40*, 339–359.

Sum4all. (n.d.). *Global Tracking Framework 2.0.* https://www.sum4all.org/global-tracking-framework.

Taylor, M., & Thompson, S. (2019). An analysis of active transport in Melbourne: Baseline activity for assessment of low carbon mobility interventions. *Urban Policy and Research, 37*(1), 62–81.

Urban Mobility Index.here. (2018). *See your city in motion and unlock the potential of how we live and move.* https://urbanmobilityindex.here.com/.

Vandycke, N. L., Care Baptista Viegas, J. M., & Morales Sarriera, J. (2019). *Global roadmap of action toward sustainable mobility* (No. 144544, pp. 1–100). The World Bank.

van de Coevering, P., Maat, K., Kroesen, M., & van Wee, B. (2016). Causal effects of built environment characteristics on travel behaviour: A longitudinal approach. *European Journal of Transport and Infrastructure Research, 16*(4), 674–697.

Veenhoven, R. (1993). *Happiness in nations: Subjective appreciation of life in 56 nations 1946–1992.* Erasmus University Rotterdam.

Veenhoven, R. (2008). Healthy happiness: Effects of happiness on physical health and the consequences for preventive health care. *Journal of Happiness Studies, 9*(3), 449–469.

Weckroth, M. T. K. (2012). *Subjective wellbeing and happiness inside a city region: Urban villages and gated communities.* (Presentation). Regional Studies Association Global Conference 2012. https://helda.helsinki.fi/bitstream/handle/10138/37353/full_paper_weckroth.pdf?sequence=2.

Weijs-Perrée, M., van den Berg, P., Arentze, T., & Kemperman, A. (2015). Factors influencing social satisfaction and loneliness: A path analysis. *Journal of Transport Geography, 45*, 24–31.

White, M. P., Alcock, I., Wheeler, B. W., & Depledge, M. H. (2013). Would you be happier living in a greener urban area? A fixed-effects analysis of panel data. *Psychological Science, 24*(6), 920–928.

White, R. L., Babic, M. J., Parker, P. D., Lubans, D. R., Astell-Burt, T., & Lonsdale, C. (2017). Domain-specific physical activity and mental health: A meta-analysis. *American Journal of Preventive Medicine, 52*(5), 653–666.

WHR. (2020). *World Happiness Report*. Sustainable Development Solutions Network. https://worldhappiness.report/.

WHR. (2021). *World Happiness Report*. Sustainable Development Solutions Network. https://happiness-report.s3.amazonaws.com/2021/WHR+21.pdf.

Willberg, E. (2019). Bike-sharing as part of urban mobility in Helsinki: A user perspective [Master's Thesis, University of Helsinki]. https://helda.helsinki.fi/bitstream/handle/10138/301581/MasterThesis_EliasWillberg.pdf?sequence=2.

Willits, F. K., & Crider, D. M. (1988, September). Health rating and life satisfaction in the later middle years. *Journal of Gerontology, 43*(5), S172–S176.

Winfield-Pfefferkorn, J. (2005). The branding of cities: Exploring city branding and the importance of brand image [Master's Thesis, Syracuse University].

Wright, L., & Montezuma, R. (2004). Reclaiming public space: The economic, environmental, and social impacts of Bogotá's transformation (Presentation). *Cities for People Conference, Walk21.* https://discovery.ucl.ac.uk/110/.

Zidansek, A. (2007). Sustainable development and happiness in nations. *Energy, 32*, 891–897.

Chapter 11
Linking Subjective Wellbeing and Pro-environmental Behaviour: A Multidimensional Approach

Christian Krekel and Alberto Prati

Abstract In recent years, policy-makers have taken steps towards acknowledging the importance of mental states when appraising citizens' wellbeing on the one side and the urgent challenge of shifting towards a more ecological society on the other. Previous work has established an encouraging positive link between these two seemingly unrelated notions, subjective wellbeing and pro-environmental behaviour. This chapter offers an overview of the progress made to date and underlines that both subjective wellbeing and pro-environmental behaviour can be structured according to different dimensions that interact in various ways. In this chapter, we empirically investigate some dimensions of both subjective wellbeing and pro-environmental behaviour that have been overlooked so far. To do so, we use newly available data collected by the French National Statistical Institute and study seven dimensions of subjective wellbeing as well as attributions of political responsibility and prioritisation about green policies. In doing so, we move beyond the often unidimensional paradigm which associates subjective wellbeing with life satisfaction and pro-environmental behaviour with consumption. Our multidimensional approach offers new insights into which dimension of subjective wellbeing is most predictive of which pro-environmental behaviour and how happy and unhappy citizens have different attitudes about environmental policies.

Keywords Subjective wellbeing · Pro-environmental behaviour · Survey · Affect · Eudaimonia · Policy preferences

C. Krekel
Department of Psychological and Behavioural Science, London School of Economics (LSE), London, UK
e-mail: c.krekel@lse.ac.uk

Centre for Economic Performance (CEP), LSE, London, UK

A. Prati (✉)
Wellbeing Research Centre, University of Oxford, Oxford, UK
e-mail: alberto.prati@hmc.ox.ac.uk

Department of Economics, Sciences Po Paris, Paris, France

S. Cloutier et al. (eds.), *Linking Sustainability and Happiness*, Community Quality-of-Life and Well-Being,
https://doi.org/10.1007/978-3-030-89559-4_11

11.1 Introduction

Anthropogenic global warming—global warming caused by humans—is one, if not the, main challenge facing humankind today. In 2019, the Intergovernmental Panel on Climate Change (IPCC) estimated that global warming is likely to reach 1.5 °C between 2030 and 2052, if it continues to increase at the current rate. This will yield significantly higher climate-related risks for natural systems, such as biodiversity loss and extinction, as well as human systems, including risks to health, livelihoods, food security, water supply, peace and security, and economic growth (Masson-Delmotte et al., 2019). These risks are unevenly distributed, yielding strong heterogeneity in who is impacted, how, when, and where. The long-term impacts of global warming, however, are likely to stay for everyone, continuing to cause long-term damages to the climate system.

To avoid overshooting 1.5 °C of global warming and its irreversible consequences to nature and humans, it is estimated that anthropogenic CO2 emissions must decline by about 45% from 2010 levels by 2030, reaching net zero at around 2050. This will require—besides technical innovation and adaptation—large-scale transformations in the ways we produce, consume, and live our daily lives. When it comes to individual and household behaviour, for example, it is estimated that changing 17 behaviours in five distinct categories can lower national US carbon emissions by about 7% (Dietz et al., 2009). Although not all pro-environmental behaviours are directly related to reductions in global warming, such behaviours almost always entail positive externalities and thereby indirectly contribute to reducing greenhouse gas emissions, thus benefiting current and future generations. There is a large consensus about the necessity of shifting towards more pro-environmental behaviours. At the same time, large-scale transformations in production and consumption require a renewed focus of policy.

Over the last two decades, the notion of subjective wellbeing as a focus (or even the ultimate goal) of policy has gained traction, at the level of national governments such as France, New Zealand, or the UK but also in international organisations such as the UN or the OECD (De Neve et al., 2020; Frijters et al., 2020; O'Donnell et al., 2020; Stiglitz et al., 2009). Can a focus on subjective wellbeing in policy-making contribute to our climate change targets, and to pro-environmental behaviour in particular?

There is reason to believe that the answer is "yes": a focus on subjective well-being in policy-making is likely to prioritise policies targeted at (mental) health, social relationships, and quality employment over private, tangible consumption.[1]

[1] It should be noted that, in this line of argument, we adopt the perspective of policy targeted at an average person living in a developed country. There can be substantial returns to subjective wellbeing from private, tangible consumption in less developed countries, or for population groups at the lower end of the income distribution in developed countries, for whom basic needs are not or just met. The relationship between subjective wellbeing and private consumption, or economic growth more generally, is hence context-dependent. Importantly, a focus on subjective wellbeing in policy-making does not run counter to the notion of economic growth; it merely reshuffles policy priorities depending on development levels. For a complete treatise of subjective wellbeing in policy-making, see Frijters & Krekel (2021).

Arguably, if subjective wellbeing promotes pro-environmental behaviour, it may be a win–win from a policy perspective, and hence a good argument to focus on subjective wellbeing in policy-making. Yet, even when moving from the macro-level of policy-making to the micro-level of individuals, there is reason to believe that subjective wellbeing may be conducive to pro-environmental behaviour. For example, there is experimental evidence that inducing positive mood and incidental happiness makes people more patient and future-oriented (Ifcher & Zarghamee, 2011; Pyone & Isen, 2018), more time-consistent (Lerner et al., 2012), less negative reciprocal (Andrade & Ariely, 2009; Drouvelis & Grosskopf, 2016) and hence more pro-social, including, potentially, towards future generations.

Although previous empirical evidence has established an encouraging link between subjective wellbeing and pro-environmental behaviour, the literature has focused mostly on a specific measure of subjective wellbeing (mostly life satisfaction) and a specific channel of pro-environmental behaviour (mostly consumption). Nevertheless, both subjective wellbeing and pro-environmental behaviour are multi-dimensional notions, the dimensions of which potentially interact in different ways. Subjective wellbeing includes overall cognitive evaluations of the present (life satisfaction), past (recalled life satisfaction), and future (expected life satisfaction) as well as relative cognitive evaluations (life satisfaction relative to others); affect; and eudaimonia. Pro-environmental behaviour includes not only responsible consumption habits but also political attitudes toward green policies. These distinctions can be crucial to tailor effective policies aimed at raising both subjective wellbeing and pro-environmental behaviour.

To provide a useful road map, this chapter starts with a selective review of the theoretical and empirical evidence on the link between subjective wellbeing and pro-environmental behaviour, an important element of sustainability and an essential component in the transition to net zero. To make progress, we move on to an empirical analysis and study the relationship between different dimensions of subjective wellbeing and pro-environmental behaviour systematically. On the one hand, we abandon the equation *subjective wellbeing* = *life satisfaction*, and look at the effects of seven different dimensions of subjective wellbeing on several types of pro-environmental behaviour. On the other hand, we move beyond consumption as the only channel of pro-environmental behaviour, and study the relationship between subjective wellbeing and heterogeneous preferences in transportation policies. To this end, we analyse recently collected data by the French National Statistical Institute and the CEPREMAP Well-Being Observatory. We match several overlapping sub-samples of a longitudinal cross-section survey and examine a representative sample of the French population from 2016 to 2019. Crucially, this newly available data set offers the possibility to study policy-relevant yet unexplored dimensions of both subjective wellbeing and pro-environmental behaviour.

We contribute to the existing literature in several ways. First, we unpack which subjective wellbeing dimension is most predictive of which pro-environmental behaviour. On top of the subjective wellbeing dimensions recently studied by Laffan (2020), we explore the predictive capacity of self-assessed life satisfaction relative to others (to capture social comparison effects), relative to oneself in the past (to capture memory effects), and relative to oneself in the future (to capture optimism). Second, we shift our focus from the behaviour of (un)happy consumers to that of (un)happy voters. To this end, we study, besides pro-environmental behaviour, attitudes towards policies related to environmental sustainability, including attributions of responsibility (who should act?) and prioritisation (what should be done?).

11.2 Literature Review

Most literature has considered how environmental quality (e.g. air pollution) affects subjective wellbeing, most often measured in terms of life satisfaction (Ambrey & Fleming, 2011; Ambrey et al., 2014; Bertram & Rehdanz, 2015; Bertram et al., in press; Ferreira et al., 2013; Kopmann & Rehdanz, 2013; Krekel & Zerrahn, 2017; Krekel et al., 2016; Levinson, 2012; Rehdanz & Maddison, 2008; Van Praag & Baarsma, 2005; von Möllendorff & Welsch, 2017). Below, we review a nascent stream of literature, which has looked at the relationship between *pro-environmental behaviour* (including attitudes, but mostly limited to consumption) and subjective wellbeing (including evaluative and affective measures, but rarely analysed jointly).

Pro-environmental behaviour and evaluative wellbeing

Most studies in this vein of research asked whether behaving in a more or less pro-environmental manner has negative or positive welfare consequences; if consequences were positive, this would imply a win–win for the environment and individuals engaging in pro-environmental behaviour.

Several studies documented the positive relationship between pro-environmental behaviour and subjective wellbeing, measured as life satisfaction or eudaimonia. In a 2018 study, for example, Schmitt et al. studied the relationship between different pro-environmental behaviours and life satisfaction in Canadian and US samples. Their results show that all but two out of 39 behaviours are predictive of life satisfaction, whereby life satisfaction seems more strongly predicted by behaviours that involve social interaction, are directly observable, and, in particular, involve more direct costs in terms of time, money, and effort. Welsch and Kühling (2018) found a positive relationship between green self-image and life satisfaction across 35 European countries, which seems to be stronger in societies with greater unanimity in pro-environmental attitudes (Binder & Blankenberg, 2017). Their results pointed towards the importance of conforming to social norms as a potential mechanism. Other potential mediators of this relationship can be the level of nature-relatedness (Nisbet et al., 2011) and nature connectedness (Martin et al., 2020), the exposure to nature (Whitburn et al., 2018), and the different individual's notion of the *good life* (Binder et al., 2020).

Pro-environmental behaviour and consumer's wellbeing

The fact that pro-environmental behaviour may pose a win–win for the environment and welfare is somewhat of a paradox, at least from the point of view of standard consumer theory. When reducing the consumption of some goods (e.g. energy), by paying more for some other goods (e.g. green products), and by paying attention to some hidden characteristics of the goods (e.g. their origin), the consumer faces *direct costs* in terms of time, money, and effort (Schmitt et al., 2018). Pro-environmental behaviour should thus, in theory, be welfare-reducing, at least for the individual engaging in it.

This paradox seems, at least partly, explicable by differential effects of different dimensions of subjective wellbeing (that is, whether one looks at affect as opposed to eudaimonia) and differential characteristics and types of pro-environmental behaviour (that is, whether one looks at consumption, comfort aspects, or the uncertainty about the ultimate usefulness of a particular behaviour). Venhoeven et al. (2013) suggested that the association between pro-environmental behaviour and the *eudemonic* dimension of subjective wellbeing is stronger than that with affect, which was also confirmed in a recent meta-analysis by Pritchard et al. (2020). The welfare consequences of pro-environmental behaviour for individuals may thus depend on exactly which dimension of subjective wellbeing one looks at and exactly which characteristic of pro-environmental behaviour is salient (that is, what kind of self-image one has, whether one thinks one can actually make a difference, or whether one feels that one conforms to a social norm, besides disutility aspects such as time, money, and effort).

There is indeed growing evidence for the importance of such behavioural factors in the relationship between pro-environmental behaviour and subjective wellbeing. These range from ego aspects such as green self-image (Welsch et al., 2021) to warm glow effects (Andreoni, 1990; Taufik et al., 2015; Welsch et al., 2021), from the misprediction of welfare consequences of spending time in nature (Nisbet & Zelenski, 2011) or consuming pro-environmentally (Welsch & Kühling, 2010) to reference point effects in consumption, either with oneself in the past or with others (Welsch & Kühling, 2011). While consuming pro-environmentally seems to be more related to experiential measures of subjective wellbeing such as happiness, ego aspects seem to be more related to evaluative measures such as life satisfaction (Welsch et al., 2021).

Pro-environmental behaviour and affective wellbeing

The studies mentioned so far have all considered how pro-environmental behaviour predicts subjective wellbeing, holding other factors constant. On the contrary, only a handful of studies have assessed this relationship in reverse, namely how subjective wellbeing predicts pro-environmental behaviour, holding other factors constant. These studies have mostly looked at the predictive power of emotions and mood, rather than of evaluative measures.

Ex-ante, the direct effect of happiness on pro-environmental choices is not clear and probably context-dependent. On the one hand, in experimental settings, inducing positive mood and incidental happiness has been shown to make people more future-oriented (Ifcher & Zarghamee, 2011; Pyone & Isen, 2018), to increase time-consistency in their choices (Lerner et al., 2012) and to reduce negative reciprocity (Riepl et al., 2016), in particular relative to other emotions (Andrade & Ariely, 2009; Drouvelis & Grosskopf, 2016). Positive mood and incidental happiness may thus render people more pro-social, including, potentially, towards future generations. However, on the other hand, happiness as a specific emotion can also lead to less deliberate cognitive processing, especially in contexts characterised by certainty (Tiedens & Linton, 2001).

Rees et al. (2015) used a lab experiment to show that, when confronted with human-caused as opposed to seemingly natural environmental damages, students report a more guilty conscience and are more likely to sign a pro-environmental petition. Indeed, Bamberg and Möser (2007) found that guilt is positively associated with perceived behavioural control, attitude, and moral norms. Bissing-Olson et al. (2016) measured guilt and pride in an experience-sampling experiment several times during the day over several days, showing that engagement in pro-environmental behaviour is positively associated with pride, negatively with guilt, and that pride predicts subsequent pro-environmental behaviour for people who hold more positive, pro-environmentally prescriptive norms. Schneider et al. (2017) found that anticipated pride as opposed to anticipated guilt from environmental action leads to more action, while Onwezen et al. (2013) found similar effects for both anticipated pride and guilt, suggesting that the effects of emotions on pro-environmental behaviours may depend, besides the specific emotion and pro-environmental behaviour, on treatment and context. Thereby, incidental emotions (that is, emotions unrelated to the decision problem at hand) may matter less than integral emotions (Ibanez et al., 2017), or in some cases even not at all (Lange & Dewitte, 2020).

Pro-environmental behaviour and relative wellbeing

Few studies looked at how *relative* satisfaction—i.e. relative to other people and other epochs—predicts different types of pro-environmental behaviour. Kaida and Kaida (2016) used a mail survey in Stockholm, Sweden, to show that expected future life satisfaction is negatively associated with current pro-environmental behaviour, potentially because more optimistic people may be less motivated to engage in behaviour change. In a previous study in Tsukuba, Japan, Kaida and Kaida (2015) showed the opposite effect for pessimism. Prati et al. (2017) found, using structural equation modelling and a student sample, that social wellbeing (measured in terms of a 33-item summed scale) at baseline predicts subsequent pro-environmental behaviour (measured in terms of energy conservation). To the best of our knowledge, no study has explored the role of perceived changes in well-being, intended as the difference between current and recalled wellbeing.

Pro-environmental behaviour and multidimensional wellbeing

The studies most closely related to ours are Wang and Kang (2018) and Laffan (2020). Wang and Kang (2018) regressed an index of pro-environmental behaviour on life satisfaction, using the China General Social Survey and estimating OLS as well as IV 2SLS models that exploit unexpected length of sunshine as an instrument. The authors found that life satisfaction has a positive effect on pro-environmental behaviour. Laffan (2020) used the 2014–2015 wave of *Natural England's Monitor of Engagement with the Natural Environment* data set and five types of pro-environmental behaviour (recycling, buying eco-friendly products, encouraging others to protect the environment, being a member of an environmental or conservation organisation, and volunteering to help care for the environment), which were collapsed into indicators of general and common *versus* uncommon pro-environmental behaviours using multiple correspondence analysis. Four different dimensions of subjective wellbeing (Dolan & Metcalfe, 2012) were then regressed on these indicators alongside controls. The author found that, while both life satisfaction and worthwhileness are positively associated with common *versus* uncommon pro-environmental behaviours in a similar manner (anxiety is strongly negatively associated), worthwhileness is a better predictor of engagement in general pro-environmental behaviour.

Our chapter builds on this previous work. On the one hand, it offers a systematic empirical comparison of the relationship between pro-environmental behaviour and subjective wellbeing in their different dimensions. On the other, it extends this analysis to a new population (French residents), different subjective wellbeing dimensions (relative, on top of absolute subjective wellbeing as well as subjective wellbeing over time, on top of current) and unexplored dimensions of pro-environmental preferences (attributions of political responsibility and policy prioritisation, on top of several types of pro-environmental behaviours).

11.3 Data and Methods

11.3.1 Research Design

For our analysis, we use recently collected data by the French National Statistical Institute. Data were collected within the *Enquête de conjoncture auprès des ménages* (CAMME), the French version of the monthly consumer confidence survey. Each month, the survey interviews a representative sample of the French population via phone. Since 2009, each November, the survey hosts an additional module on the environment, "Environmental habits and opinions." Since 2016, thanks to a joint effort between the French National Statistical Institute and the CEPREMAP Well-Being Observatory, the survey includes a quarterly module on subjective wellbeing (in March, June, September, and December). The data set is structured as a longitudinal cross-section with partially overlapping sub-samples. One third of the sample is renewed every month, so that each person is interviewed three times over three consecutive months. Therefore, since 2016, answers from the environmental module can be matched with answers from the subjective wellbeing module, either precedent or subsequent. Figure 11.1 offers a visual illustration of the timeline of these overlapping samples.

After the matching process, we end up having observations on 5,733 individuals. Each individual is observed twice (in September and November or in November and

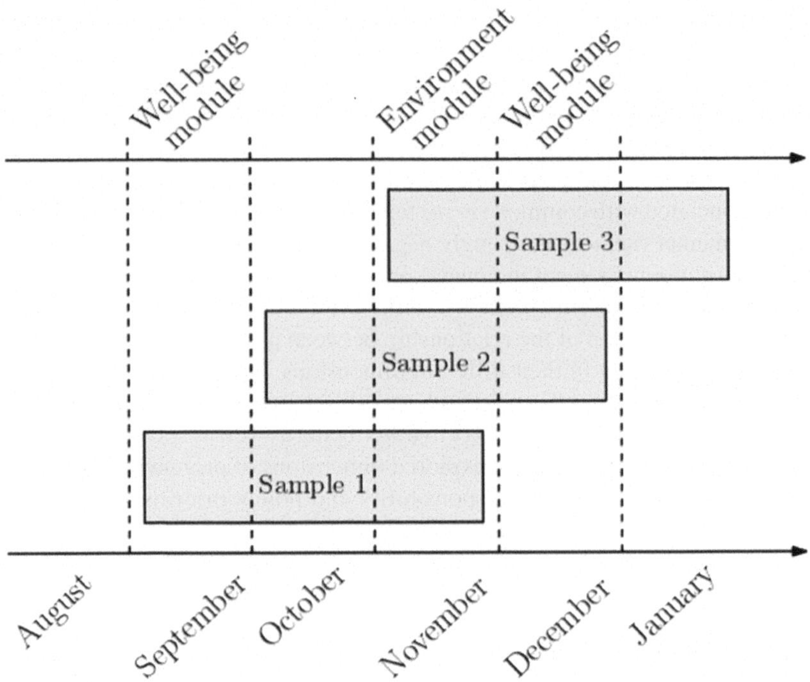

Fig. 11.1 Timeline of the survey. *Note* Each sub-sample is interviewed three months in a row. The survey hosts a module on the environment in November and a module on subjective wellbeing in September and December

December) and replies exactly once to each module. The present chapter is the first study which explores this newly available data set.[2]

11.3.2 Variables and Summary Statistics

The subjective wellbeing module asks 20 questions about different dimensions of subjective wellbeing, including evaluations, experiences, and eudaimonia. In this chapter, we will focus on seven indicators: (1) general life satisfaction, (2) life satisfaction with respect to others, (3) life satisfaction with respect to one's own future, (4) life satisfaction with respect to one's own past, (5) experienced happiness yesterday, (6) experienced depression yesterday, and (7) eudaimonia (sense of purpose in life). Questions are detailed in the Appendix (Table 11.9). Table 11.1 presents summary statistics.

[2] Data are publicly available upon request to the *French Data Archive for Social Science*. The files we used for our analysis are: lil-1309, lil-1373 lil-1176, and lil-1253.

Table 11.1 Summary statistics: subjective wellbeing

	N	mean	sd	min	max
Life satisfaction	5703	6.53	1.71	0	10
Expected life satisfaction	5447	5.75	1.99	0	10
Recalled life satisfaction	5682	6.38	1.89	0	10
Life satisfaction w.r.t others	5541	6.53	1.68	0	10
Life satisfaction w.r.t future	5436	0.79	1.50	−8	10
Life satisfaction w.r.t past	5667	0.15	1.55	−10	10
Happy yesterday	5678	6.79	2.20	0	10
Not depressed yesterday	5684	7.93	2.72	0	10
Eudaimonia	5573	7.01	1.96	0	10

Note $N = 18{,}435$. The mean life satisfaction in our sample is 6.53 (with values ranging from 0 to 10), obtained from 5,703 observations. The standard deviation is 1.71

The first indicator—general life satisfaction—is the gold standard to assess evaluative subjective wellbeing. On the contrary, questions on relative, expected, and recalled life satisfaction are much rarer in national and international surveys, but they can be highly informative about relative inter-personal and intra-personal comparisons. We observe in our data that the vast majority of people think that they are more satisfied with their lives than others (Fig. 11.2).

This finding is in line with the well-known *above-average bias*: most people think they are better than average on virtually every desirable trait (Gilovich, 2008), including happiness (Lykken & Tellegen, 1996). Although several studies have shown the importance of relative comparisons for subjective wellbeing, it is yet unclear to what extent perceived relative subjective wellbeing is associated with (different) behaviour(s). For the purpose of obtaining an indicator of relative life satisfaction with respect to one's own future, we calculate the difference (in scale-point units)

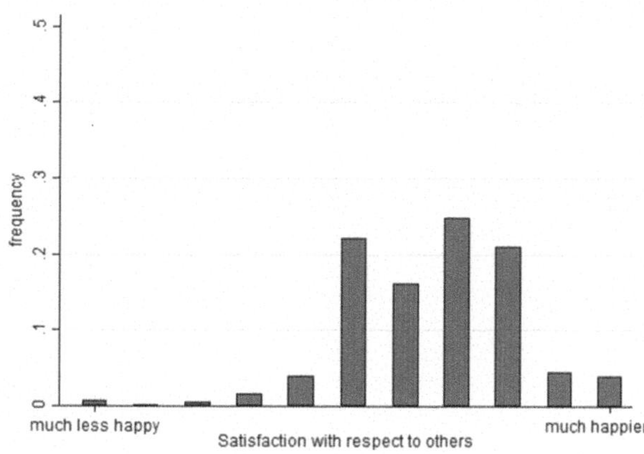

Fig. 11.2 Distribution of perceived *inter*-personal satisfactions. Satisfaction w.r.t. others

between current and expected future life satisfaction of a respondent. We do the same for recalled past life satisfaction.

These variables then reflect satisfaction with respect to the future and satisfaction with respect to the past, respectively. They both contain information on the perceived life satisfaction trajectory (ascending or descending) of a respondent, but along different directions of time. Figure 11.3a, b sketch the distributions of these indicators: both are positive-skewed, suggesting that expected future and recalled past life satisfaction are not random, and hence can have behavioural implications.

Questions on recently experienced happiness and depression are typically used as complementary measures. For ease of interpretation, we reverse code "depression yesterday" so that for all variables higher values correspond to higher subjective wellbeing. The last question is aimed at measuring eudaimonia, a dimension that, in spite of its theoretical importance, has received scant empirical attention so far. All

Fig. 11.3 **a** Distribution of perceived *intra*-personal satisfactions (future). Satisfaction w.r.t. future. **b** Distribution of perceived *intra*-personal satisfactions (past). Satisfaction w.r.t. past

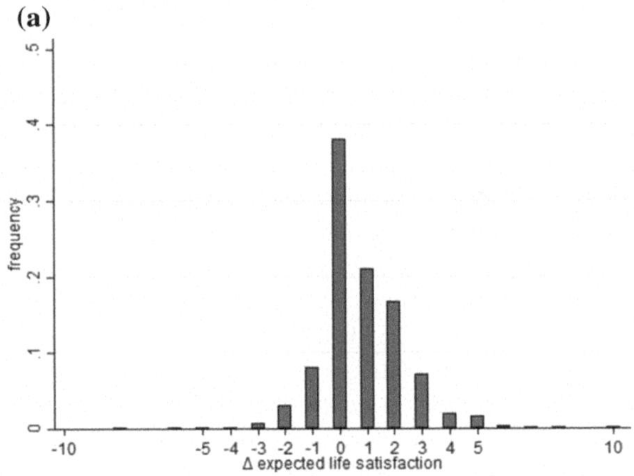

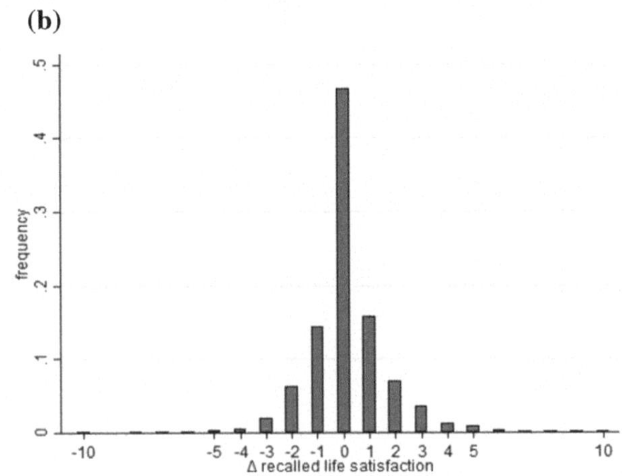

questions are answered on a 0-to-10 scale. Indicators (3) and (4) are computed as differences and thus take values within the set (−10, 10).

The environmental module is made of 20 questions, too. It seeks to document the spread of different pro-environmental behaviours and opinions about environmental policies. We focus on eight behaviours, specifically: (1) the propensity to buy organic food or (2) eco-labelled products; (3) the attention devoted to purchasing local food, (4) local products, (5) or low-waste products; (6) the propensity to switch off unused electrical devices and (7) to lower the heat or the air conditioning; and (8) the means of transport which is usually adopted to go shopping. The first two questions are answered as yes–no, while questions (3) to (7) are reported on an ordered 1-to-4 scale (*never, sometimes, often, always*). Finally, the mode of transport is reported on a list of several options, which we dichotomise as "car/motorbike" *versus* not. Questions are reported in the Appendix (Table 11.10). Table 11.2 presents summary statistics.

On top of their pro-environmental behaviours, people are also asked about their opinions on the political responsibility to protect the environment and about which factor would encourage them the most to use green transportation. Section 4.2 describes these variables in detail.

Finally, the data set contains a set of standard socio-demographic characteristics of respondents, including age, sex, education level, job status (employed, unemployed, inactive, student, retired), the category of a respondent's occupation (managers, intermediate, clerks, workers), income, the living situation (whether living as a couple), the region of residence, and the size of the unit of residence. Table 11.3 presents summary statistics.

Table 11.2 Summary statistics: pro-environmental behaviours

	N	mean	sd	min	max
Organic	5661	0.60	0.49	0	1
Eco-label	5165	0.59	0.49	0	1
Local food	5663	2.55	1.13	0	4
Local others	5656	2.18	1.07	0	4
No waste	5660	2.16	1.04	0	4
Switch off	5665	2.64	1.13	0	4
AC/heat	5639	2.94	1.06	0	4
Car to shop	5539	0.17	0.38	0	1

Note $N = 18{,}435$. 60% of 5661 respondents went shopping to an organic store during the last month. The standard deviation is 0.49

Table 11.3 Summary statistics: Socio-demographic characteristics

	N	mean	sd	min	max
Age	5733	57.2	15.8	18	97
Male	5733	0.48	0.49	0	1
No diploma	5733	0.29	0.45	0	1
High school	5733	0.18	0.38	0	1
Undergraduate	5733	0.22	0.41	0	1
Graduate	5733	0.13	0.34	0	1
Log (income)	4888	7.79	0.59	5.15	9.74

11.3.3 Models

We seek to answer the following question: on average and everything else equal, are people who report higher subjective wellbeing more likely to adopt a particular pro-environmental behaviour? To answer this question, we adopt a multiple regression framework in which we regress each of the eight types of pro-environmental behaviour on a different measure of subjective wellbeing, controlling for the socio-demographic characteristics of respondents. For ease of interpretation of marginal effects, we use a linear model, but results are similar if a non-linear model is used (probit or ordered probit, depending on the outcome).

We estimate a regression of the following form:

$$y_{it} = \alpha x_{it} + Z_{it}\beta + \gamma \tau_t + \varepsilon_{it} \tag{11.1}$$

where y_{it} is the relative frequency with which respondent i adopts pro-environmental behaviour y in year t; x_{it} is an individual-specific measure of subjective wellbeing; Z_{it} is a set of socio-demographic characteristics of respondents as controls; τ_t captures year fixed effects; ε_{it} is the normally-distributed error term. In our analysis, we systematically change the content of y_{it} and x_{it} to study the relationship between different dimensions of both pro-environmental behaviour and subjective wellbeing.

By controlling for socio-demographic characteristics, we aim to look at the predictive effect of subjective wellbeing on pro-environmental behaviour over and beyond these characteristics. For example, by controlling for income, we make sure that any potential, positive association between subjective wellbeing and pro-environmental behaviour is not merely driven by the fact that respondents who have a higher income can financially afford to consume more pro-environmentally, both of which is positively associated with subjective wellbeing. In other words, we are estimating the predictive power of subjective wellbeing on pro-environmental behaviour net of such socio-demographic differences.

11.4 Results and Discussion

11.4.1 Subjective Wellbeing and Pro-environmental Behaviour

Table 11.4 presents estimates where x_{it} is life satisfaction. The eight columns refer to different specifications where only the dependent variable y_{it} changes.

In line with previous studies (Kaida & Kaida, 2016; Laffan, 2020; Prati et al., 2017; Wang & Kang, 2018), life satisfaction predicts the adoption of several pro-environmental behaviours, and importantly, this holds net of socio-demographic differences. A one-point increase in life satisfaction is associated with a 2% increase in the likelihood of buying organic or eco-labelled products (columns 1 and 2). Higher life satisfaction is also associated with higher attention paid at purchasing local food and low-waste products (columns 3 and 5). However, we find no significant association with energy consumption or transportation habits, which seem to be predicted mostly by age and sex (energy) and income and living as a couple (transportation).

Before moving on to the different subjective wellbeing dimensions, let us first discuss associations between socio-demographic characteristics and different pro-environmental behaviours. Age is generally associated with more pro-environmental behaviours (in a concave fashion) and so is education (in a convex fashion). Men tend to be less attentive to the environment than women are, while income intervenes in opposite ways: richer people are more prone to buying eco-responsible products (which are typically more expensive) but are less attentive to saving energy.[3] Finally, the dummy variables for survey years suggest an encouraging trend: in 2019, several pro-environmental behaviours were significantly more common than three years before.

Tables 11.5, 11.6 and 11.7 present regression results where the regressor x_{it} is systematically changed. For the sake of parsimony, in these tables, we omit the estimated coefficients of the socio-demographic controls (they are quite similar to the ones presented in Table 11.4).

Table 11.5 studies the predictive power of relative comparisons and temporal dynamics in life satisfaction. Satisfaction with respect to others (panel 1) is positively correlated with some behaviours, including AC/heat saving, but its overall predictive power is more limited than life satisfaction in general. One might have expected that the perception of doing better in life than others might incentivise people to take on more responsibility towards the environment. Contrary to our expectations, the indicators of perceived life satisfaction with respect to one's own past (panel 2) and with respect to one's own future (panel 3) perform poorly. The finding that expectations about life satisfaction in the future (often used as a proxy for optimism) do not feature negatively for current levels of pro-environmental behaviour stands in contrast to Kaida and Kaida (2016), who find that a more optimistic outlook depresses

[3] Because of space constraints, estimates associated with gender and education level are not displayed in the table.

Table 11.4 Regression of several pro-environmental behaviours on life satisfaction

	(1)	(2)	(3)	(4)	(5)	(6)	(7)	(8)
	Organic	Ecolabel	Local food	Local others	No waste	Switch off	AC/heat	Car to shop
Life satisfaction	0.0130***	0.0190***	0.0452***	0.0060	0.0275***	0.0123	0.0103	0.0032
	(0.0047)	(0.0047)	(0.0112)	(0.0107)	(0.0103)	(0.0111)	(0.0108)	(0.0034)
Age	0.0080**	0.0147***	0.0233***	0.0379***	0.0167**	0.0286***	0.0190***	−0.0040
	(0.0032)	(0.0033)	(0.0076)	(0.0068)	(0.0069)	(0.0079)	(0.0072)	(0.0025)
Age^2	−0.0001**	−0.0001***	−0.0002**	−0.0002***	−0.0001**	−0.0002***	−0.0002***	0.0000*
	(0.0000)	(0.0000)	(0.0001)	(0.0001)	(0.0001)	(0.0001)	(0.0001)	(0.0000)
Log (income)	0.0856***	0.0629***	0.1184***	0.0580	−0.1123***	−0.1555***	0.0107	−0.0598***
	(0.0188)	(0.0187)	(0.0440)	(0.0429)	(0.0392)	(0.0444)	(0.0424)	(0.0143)
Year 2017	0.0302	−0.0020	0.0743	−0.0475	0.0311	0.1174**	0.0961**	−0.0124
	(0.0208)	(0.0208)	(0.0490)	(0.0454)	(0.0453)	(0.0488)	(0.0461)	(0.0144)
Year 2018	0.0709***	0.0012	0.1206**	−0.0814*	0.0923**	0.0623	0.1246***	0.0005
	(0.0208)	(0.0212)	(0.0491)	(0.0458)	(0.0461)	(0.0487)	(0.0462)	(0.0146)
Year 2019	0.0963***	0.0374*	0.1981***	−0.1205***	0.1685***	0.0018	0.1245***	0.0138
	(0.0202)	(0.0205)	(0.0473)	(0.0444)	(0.0447)	(0.0472)	(0.0454)	(0.0150)
N	4231	4231	4231	4231	4231	4231	4231	4231
Adjusted R^2	0.091	0.060	0.058	0.064	0.024	0.022	0.020	0.191

Note Standard errors in parentheses. $^{*}p < 0.10$, $^{**}p < 0.05$, $^{***}p < 0.01$. The regressions include a constant and controls for job characteristics (type and position), size of the unit of residence (rural, small city, big city, Paris), and region fixed effects

Table 11.5 Regression of several pro-environmental behaviours on relative satisfaction

	(1)	(2)	(3)	(4)	(5)	(6)	(7)	(8)
	Organic	Ecolabel	Local food	Local others	No waste	Switch off	AC/heat	Car to shop
Satisfaction w.r.t others	0.0070	0.0140***	0.0364***	0.0098	0.0230**	0.0181	0.0271**	0.0031
	(0.0048)	(0.0049)	(0.0113)	(0.0110)	(0.0106)	(0.0114)	(0.0112)	(0.0035)
N	4136	4136	4136	4136	4136	4136	4136	4136
Adjusted R^2	0.090	0.058	0.058	0.066	0.022	0.023	0.019	0.193
Satisfaction w.r.t past	0.0095**	0.0061	0.0065	−0.0074	0.0088	0.0108	−0.0151	−0.0057*
	(0.0048)	(0.0048)	(0.0117)	(0.0110)	(0.0106)	(0.0115)	(0.0107)	(0.0034)
N	4211	4211	4211	4211	4211	4211	4211	4211
Adjusted R^2	0.090	0.056	0.054	0.064	0.022	0.022	0.020	0.191
Satisfaction w.r.t future	0.0017	0.0042	0.0092	0.0161	0.0034	0.0026	−0.0014	−0.0019
	(0.0049)	(0.0050)	(0.0120)	(0.0112)	(0.0112)	(0.0120)	(0.0111)	(0.0035)
N	4077	4077	4077	4077	4077	4077	4077	4077
Adjusted R^2	0.089	0.054	0.056	0.066	0.023	0.023	0.017	0.195
Controls	Yes	Yes	Yes	Yes	Yes	Yes	Yes	Yes
Time f.e	Yes	Yes	Yes	Yes	Yes	Yes	Yes	Yes
Regional f.e	Yes	Yes	Yes	Yes	Yes	Yes	Yes	Yes

Note Standard errors in parentheses $^{*}p < 0.10$, $^{**}p < 0.05$, $^{***}p < 0.01$. The regressions include a constant and controls for age, age^2, sex, education, marital status, log(income), job characteristics (type and position), size of the unit of residence (rural, small city, big city, Paris), year fixed effects, and region fixed effects

Table 11.6 Regression of several pro-environmental behaviours on experiential subjective wellbeing

	(1)	(2)	(3)	(4)	(5)	(6)	(7)	(8)
	Organic	Ecolabel	Local food	Local others	No waste	Switch off	AC/heat	Car to shop
Happy	0.0089***	0.0085**	0.0272***	0.0076	0.0222***	0.0100	0.0109	0.0004
	(0.0034)	(0.0034)	(0.0081)	(0.0077)	(0.0076)	(0.0082)	(0.0078)	(0.0026)
N	4223	4223	4223	4223	4223	4223	4223	4223
Adjusted R^2	0.090	0.057	0.056	0.065	0.024	0.022	0.020	0.190
Depressed	−0.0004	−0.0037	0.0026	−0.0155**	−0.0041	−0.0079	−0.0048	−0.0019
	(0.0027)	(0.0027)	(0.0064)	(0.0062)	(0.0060)	(0.0065)	(0.0061)	(0.0020)
N	4223	4223	4223	4223	4223	4223	4223	4223
Adjusted R^2	0.088	0.056	0.054	0.066	0.022	0.022	0.019	0.192
Controls	Yes	Yes	Yes	Yes	Yes	Yes	Yes	Yes
Time f.e	Yes	Yes	Yes	Yes	Yes	Yes	Yes	Yes
Regional f.e	Yes	Yes	Yes	Yes	Yes	Yes	Yes	Yes

Note standard errors in parentheses $^{*}p < 0.10$, $^{**}p < 0.05$, $^{***}p < 0.01$. The regressions include a constant and controls for age, age^2, sex, education, marital status, log(income), job characteristics (type and position), size of the unit of residence (rural, small city, big city, Paris), year fixed effects, and region fixed effects

Table 11.7 Regression of several pro-environmental behaviours on eudaimonia

	(1)	(2)	(3)	(4)	(5)	(6)	(7)	(8)
	Organic	Ecolabel	Local food	Local others	No waste	Switch off	AC/heat	Car to shop
Eudaimonia	0.0121***	0.0186***	0.0563***	0.0307***	0.0457***	0.0225**	0.0368***	0.0049
	(0.0040)	(0.0040)	(0.0094)	(0.0089)	(0.0086)	(0.0095)	(0.0092)	(0.0030)
N	4163	4163	4163	4163	4163	4163	4163	4163
Adjusted R^2	0.091	0.060	0.063	0.068	0.030	0.023	0.025	0.194
Controls	Yes	Yes	Yes	Yes	Yes	Yes	Yes	Yes
Time f.e	Yes	Yes	Yes	Yes	Yes	Yes	Yes	Yes
Regional f.e	Yes	Yes	Yes	Yes	Yes	Yes	Yes	Yes

Note Standard errors in parentheses $^{*}p < 0.10$, $^{**}p < 0.05$, $^{***}p < 0.01$. The regressions include a constant and controls for age, age^2, sex, education, marital status, log(income), job characteristics (type and position), size of the unit of residence (rural, small city, big city, Paris), year fixed effects, and region fixed effects pro-environmentalism in the present, and by the same token, in contrast to Kaida and Kaida (2015) who find the opposite for pessimism. These differences may, however, also be due to different operationalisations and scales used in these studies.

Table 11.6 looks at experiential subjective wellbeing. Although variations in happiness levels significantly predict variations in some pro-environmental behaviours (panel 1), the magnitude of the effect of a one-point variation is systematically smaller than for life satisfaction (the coefficients are comparable since both variables are measured on a 0-to-10 scale). Feelings of depression (panel 2) do not predict pro-environmental behaviour. Overall, none of the measures of experiential subjective wellbeing outperforms life satisfaction, which may point towards green self-image or life-style as a major mediator in the relationship between pro-environmental behaviour and wellbeing (Binder & Blankenberg, 2017; Welsch et al., 2021). Given our research design, it may, however, be more difficult to capture more short-term, warm-glow effects from consuming pro-environmentally as documented elsewhere (Andreoni, 1990; Taufik et al., 2015; Welsch et al., 2021).

Eudaimonia, instead, turns out to be the best predictor of pro-environmental behaviour, in line with Venhoeven et al. (2013), Pritchard et al. (2020), and recent findings by Laffan (2020). Table 11.7 presents the estimated coefficients. The adjusted R^2 shows that the explanatory power of a model including eudaimonia is systematically higher than the preceding models. Variations in eudaimonia are positively associated not only with sustainable consumption choices but also with saving energy. The magnitude of the marginal effects are, however, small: for instance, a movement all the way up the scale from 0 to 10 is associated with an upward shift of only 0.03 standard deviations in the variable AC/heat.

A null result which is worth some discussion is the inability of any of the aforementioned subjective wellbeing dimensions to predict transportation habits. On the one hand, commuting to work depends on factors which are often hard to change. On the other, we could expect that the choice of *how* to go shopping is relatively flexible in non-rural areas. In spite of this conjecture, we find no evidence of a statistical association between subjective wellbeing measures and going shopping by car or motorbike. This result suggests that one should be cautious when aggregating green transportation choices into a general index of pro-environmental behaviour.

Although the specification of our model suggests a flow from subjective wellbeing to pro-environmental behaviour, this analysis cannot make any claims of causality. It lacks an identification strategy which would allow to precisely isolate only one direction of the causal channel. Future research should focus on this aspect, which is vital for policymaking aimed at raising both pro-environmental behaviour and subjective wellbeing.

11.4.2 Subjective Wellbeing and Sustainable Policies

In the previous section, we documented a positive association between some dimensions of subjective wellbeing (in particular, general life satisfaction and eudaimonia) and different types of pro-environmental behaviours (in particular, responsible consumption habits). The analysis of pro-environmental behaviour places the consumer at the centre of the ecological transition. Yet, governments play a substantial role as well and, in a democracy, their political choices will be largely determined by citizens' preferences. In this section, we adopt a more policy-oriented perspective and explore how subjective wellbeing relates to preferences for different types of sustainable policies. First, we document that French citizens consider public authorities as the main actor for environmental protection. We then move on to a case study of transportation policies, and investigate how opinions of happy and unhappy citizens can diverge. In this section, we limit our analysis to one subjective wellbeing dimension (general life satisfaction).

Table 11.8 compares the distribution of answers to the question "Who do you think should act primarily for the protection of the environment?", conditional on life satisfaction being very low (1–4), medium–low (5–6), medium–high (7–8), or very high (9–10). The last column of the table confirms the perceived importance of public policies: most people (52.3%) think that it is primarily public authorities' responsibility to act on environmental protection, ahead of households (20.0%) and companies (27.7%).

The cross-tabulation allows us to investigate if opinions about political responsibilities on environmental protection change across life satisfaction levels. *Ex-ante* differential opinions between happy and unhappy citizens, if any, are non-trivial. On the one hand, public opinion research has shown that happier people tend to have higher trust in institutions, so that they might think that *public authorities* should be the main actor of a green transition. On the other, psychological research has often found that happier people tend to have a stronger internal locus of control (i.e. they

Table 11.8 Responsibilities about environmental protection and life satisfaction

Life satisfaction					
	Very low	Medium–low	Medium–high	Very high	Total
Public authorities (%)	55.1	54.0	51.0	50.6	52.3
Households (%)	18.1	17.6	21.6	22.7	20.0
Companies (%)	26.8	28.4	27.5	26.6	27.7
N	481	1882	2753	409	5525

Note Cross-tabulation of the responses to the question "Who do you think should act primarily for the protection of the environment?" and "Overall, how satisfied are you with your life nowadays?" Very low = 0–4; medium–low = 5–6; medium–high = 7–8; very high = 9–10. Among the 481 people reporting very low life satisfaction, 55.1% answer that public authorities are the ones that should act primarily for the protection of the environment. Among the 409 people reporting very high life satisfaction, 50.6% give the same answer believe they have control over events, as opposed to events being imposed on them), which implies that they might feel that *households* are the ones who have the power to change things. Table 11.8 offers (slight) descriptive support to the second hypothesis. As life satisfaction increases, the table shows a progressive shift of attributed responsibilities from public authorities to households. Specifically, 23% of people with high general life satisfaction thinks that it is primarily citizens' responsibility to take action to protect the environment, while only 18% of people with low current and expected life satisfaction hold the same opinion. Instead, the fraction of citizens pointing out at companies' responsibilities is stable, around 1/4 of the sample.

If most people count on public authorities, which policies should be enacted? For illustrative purposes, we now zoom in on transportation policies, which are particularly interesting for two reasons: (1) In May 2020 (thus, after our survey), the French government introduced a Mobility Plan which covers expenses related to sustainable commuting up to 400 Euro. For instance, the plan covers the purchase of a bike or an electric scooter. Who will benefit more from it? (2) It is well-known that people who usually walk or bike in their daily commutes are more satisfied with their lives (Lancée et al., 2017; Olsson et al., 2012; Wild & Woodward, 2019), but the direction of causality is unclear. In this respect, looking at the propensity of non-cyclists to endorse pro-cyclists policies can help disentangle causality.

We study answers to the following question: "As for your daily travels (work, study, leisure, shopping), which factor would encourage you the most to decrease the use of your car / motorbike / scooter?" Fig. 11.4 presents the distributions of the answers. When asked what is the best policy to diminish the usage of cars, about one third of respondents suggest enhancing public transport and another third says that none of the proposed measures would be effective.

Figure 11.5 shows that the minority of respondents (10%) who suggest improving bike infrastructures are significantly more satisfied with their lives than the rest of the

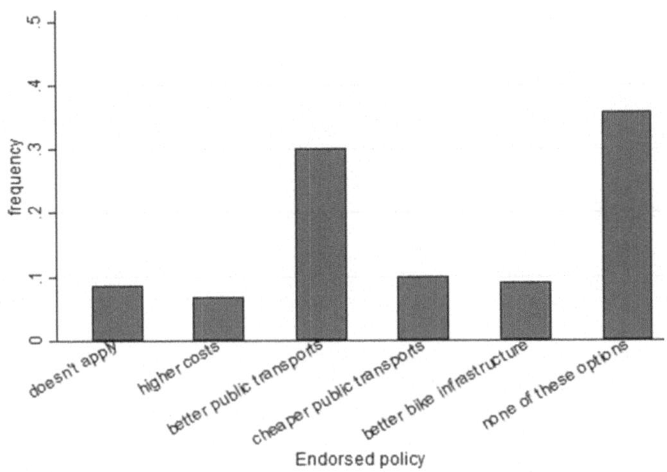

Fig. 11.4 Distribution of endorsed policies

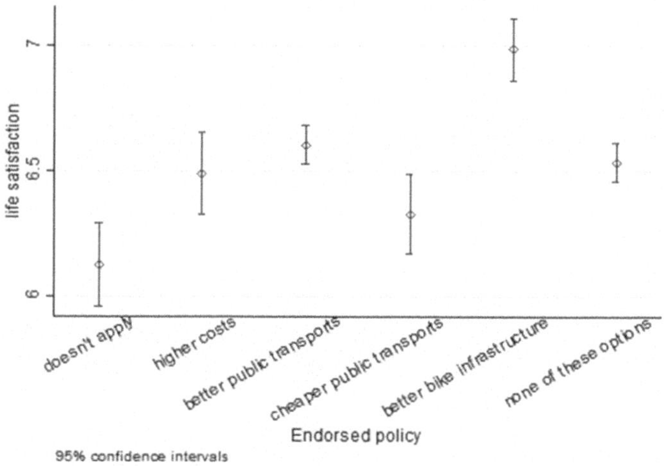

Fig. 11.5 Life satisfaction, by endorsed policy

population. This could be entirely driven by the sub-population of cyclists, who, at the same time, is more satisfied with their life and wishes to improve their infrastructure. Yet, this difference in life satisfaction also holds among people who report that they do not use bikes for their daily commuting or to go shopping. Of course, respondents' age and the size of the urban area are likely to be correlated both with their life satisfaction and with their propensity to use a bike. We can control for these factors by regressing life satisfaction on the standard set of socio-demographic characteristics and calculating the residuals. Thereby, the vector of residual life satisfaction is, by construction, orthogonal to these characteristics of respondents. When we compare conditional average residual life satisfaction (Fig. 11.6), we notice that people who ask to primarily improve bike infrastructure in order to reduce the usage of car are still more satisfied with their lives, everything else equal, even if *they do not usually use a bike.* This conditional correlation suggests two things: (1) policies which endorse bike usage are more likely to benefit people who are already more satisfied with their lives and (2) people who are more satisfied with their lives have a higher propensity for bike usage, everything else equal.

A word of caution is needed. This correlational analysis cannot make causal claims. Our analysis is exploratory in nature and can only offer suggestive evidence. However, it offers descriptive evidence on differential preferences for sustainable policies about a topic – public transport – which contributes to over 60% of nitrogen dioxide emissions and 14% of PM_{10} pollution in France.[4]

[4] Official estimates by the French governments are published here: https://www.ecologie.gouv.fr/pollution-lair-origines-situation-et-impacts.

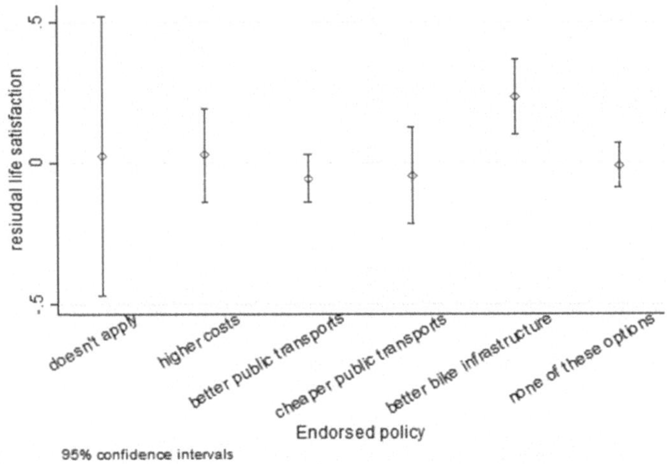

Fig. 11.6 Residual life satisfaction, by endorsed policy

11.5 Conclusion

This chapter contributed to unpacking the relationship between pro-environmental behaviour and subjective wellbeing. It started with a critical review of the theoretical and empirical evidence on whether, how, when, and why subjective wellbeing may be linked to pro-environmental behaviour, and vice versa. Both pro-environmental behaviour and subjective wellbeing are complex, rich notions which can be structured according to different dimensions. We showed that the literature has focused mostly on how a specific dimension of subjective wellbeing (mostly life satisfaction) correlates with general indices of pro-environmental behaviours or single behaviours in isolation. Some important subjective wellbeing dimensions (such as eudaimonia or inter-temporal and inter-personal life satisfaction) and some relevant individual attitudes (policy priorities) have received scant or no attention so far. Thanks to a newly available data set, surveying a representative sample of the French population over four years, this chapter conducted a systematic comparison of these dimensions.

Our results confirm the well-established positive correlation between pro-environmental behaviour and subjective wellbeing (conditional on a range of socio-demographic characteristics). Our multidimensional approach offers insights into which dimensions of subjective wellbeing are most predictive of which pro-environmental behaviours. While shopping habits seem to be strongly subjective-wellbeing-dependent, transportation habits turn out to be hardly predictable by any subjective wellbeing dimension (even if the transportation habit is not measured as commuting to work). Contrary to our expectations, inter-temporal and inter-personal measures of life satisfaction are *not* powerful predictors of green habits, a result that stands in contrast to Kaida and Kaida (2015, 2016). Instead, a highly perceived sense of purpose in life is the best predictor of many pro-environmental behaviours (Laffan, 2020).

In this chapter, we also attempted to predict asymmetric welfare effects of a sustainable policy recently introduced by the French government. We show that this policy (which financially incentivises commuting by bike or electric scooter) is more likely to benefit people who *already* report a high level of life satisfaction. Indeed, we uncover a positive correlation between propensity to endorse pro-bike policies and life satisfaction, even among daily car users and conditional on socio-demographic characteristics. This exploratory analysis invites policy makers to take into consideration differential preferences among groups of happy and unhappy citizens. Moreover, it helps to better understand the specific mechanisms for why cyclists tend to be relatively happier: they seem to be cyclists because they are happier, and not only the other way round.

Appendix

See Appendix Tables 11.9 and 11.10.

Table 11.9 Wellbeing indicators

Indicator	Question
General life satisfaction	Overall, how satisfied are you with your life nowadays?
Expected life satisfaction	When you think at next years, are you satisfied with this perspective?
Recalled life satisfaction	And when you think at last year, where would you place yourself on a scale from 0 to 10?
Life satisfaction with respect to others	And when you compare yourself to other people living in France in general, how do you feel on a scale from 0 to 10?
Happiness yesterday	Yesterday, did you feel happy?
Depression yesterday	Yesterday, did you feel depressed?
Eudaimonia	Do you feel that what you do in your life is meaningful, worthwhile?

Table 11.10 Pro-environmental behaviours

Indicator	Question
Organic	During the last month, have you (or a member of your household) purchased either in an organic shop or organic products from the supermarket?
Eco-label	During the last month, have you (or a member of your household) purchased one or more eco-labelled product(s)?
Local food	When you buy food items, do you pay attention to the distance covered for their transportation (their geographical origin)?
Local others	When you buy non-food items, do you pay attention to the place where they were produced (their geographical origin)?
No waste	When you buy some products, do you pay attention to how much waste it implies?
Switch off	At home, do you sometimes switch off the electronic devices which are in stand by mode?
AC/heat	At home, do you sometimes lower the heating or the air conditioning to save your energy consumption?
Car to shop	Typically, which of the following means of transport do you use to go shopping?

References

Ambrey, C., & Fleming, C. (2011). Valuing scenic amenity using life satisfaction data. *Ecological Economics, 72*, 106–115.

Ambrey, C., Fleming, C., & Chan, A. C. (2014). Estimating the cost of air pollution in South East Queensland: An application of the life satisfaction non-market valuation approach. *Ecological Economics, 97*, 172–181.

Andrade, E., & Ariely, D. (2009). The enduring impact of transient emotions on decision making. *Organizational Behavior and Human Decision Processes, 109*(1), 1–8.

Andreoni, J. (1990). Impure altruism and donations to public goods: A theory of warm glow giving. *Economic Journal, 100*(401), 464–477.

Bamberg, S., & Möser, G. (2007). Twenty years after Hines, Hungerford, and Tomera: A new meta-analysis of psycho-social determinants of pro-environmental behaviour. *Journal of Environmental Psychology, 27*(1), 14–25.

Bertram, C., Goebel, J., Krekel, C., & Rehdanz, K. (in press). Urban land use fragmentation and human wellbeing. *Land Economics*.

Bertram, C., & Rehdanz, K. (2015). The role of urban green space for human well-being. *Ecological Economics, 120*, 139–152.

Binder, M., & Blankenberg, A. K. (2017). Green lifestyles and subjective well-being: More about self-image than actual behavior? *Journal of Economic Behavior & Organization, 137*, 304–323.

Binder, M., Blankenberg, A. K., & Guardiola, J. (2020). Does it have to be a sacrifice? Different notions of the good life, pro-environmental behavior and their heterogeneous impact on well-being. *Ecological Economics, 167*, 106448.

Bissing-Olson, M., Fielding, K., & Iyer, A. (2016). Experiences of pride, not guilt, predict pro-environmental behavior when pro-environmental descriptive norms are more positive. *Journal of Environmental Psychology, 45*, 145–153.

Dietz, T., Gardner, G., Gilligan, J., Stern, P., & Vandenbergh, M. (2009). Household actions can provide a behavioral wedge to rapidly reduce US carbon emissions. *PNAS, 106*(44), 18452–18456.

Dolan, P., & Metcalfe, R. (2012). Measuring subjective wellbeing: Recommendations on measures for use by national governments. *Journal of Social Policy, 41*(2), 409–427.

Drouvelis, M., & Grosskopf, B. (2016). The effects of induced emotions on pro-social behaviour. *Journal of Public Economics, 134*, 1–8.

Ferreira, S., Akay, A., Brereton, F., Cuñado, J. C., Martinsson, P., Moro, M., & Nigal, T. (2013). Life satisfaction and air quality in Europe. *Ecological Economics, 88*, 1–10.

Frijters, P., Clark, A., Krekel, C., & Layard, R. (2020). A happy choice: Wellbeing as the goal of government. *Behavioural Public Policy, 4*(2), 126–165.

Frijters, P., & Krekel, C. (2021). *A handbook for wellbeing policy-making: History, theory, measurement, implementation, and examples*. Oxford University Press.

Gilovich, T. (2008). *How we know what isn't so*. Simon and Schuster.

Ibanez, L., Moureau, N., & Roussel, S. (2017). How do incidental emotions impact pro-environmental behavior? Evidence from the dictator game. *Journal of Behavioral and Experimental Economics, 66*, 150–155.

Ifcher, J., & Zarghamee, H. (2011). Happiness and time preference: The effect of positive affect in a random-assignment experiment. *American Economic Review, 101*(7), 3109–3129.

Kaida, N., & Kaida, K. (2015). Facilitating pro-environmental behavior: The role of pessimism and anthropocentric environmental values. *Social Indicators Research, 126*, 1243–1260.

Kaida, N., & Kaida, K. (2016). Pro-environmental behavior correlates with present and future subjective well-being. *Environment, Development and Sustainability, 18*(1), 111–127.

Kopmann, A., & Rehdanz, K. (2013). A human well-being approach for assessing the value of natural land areas. *Ecological Economics, 93*, 20–33.

Krekel, C., Kolbe, J., & Wüstemann, H. (2016). The greener, the happier? The effect of urban land use on residential well-being. *Ecological Economics, 121*, 117–127.

Krekel, C., Rechlitz, J., Rode, J., & Zerrahn, A. (2020). Quantifying the externalities of renewable energy plants using wellbeing data: The case of biogas (Discussion paper No. 1738). Centre for Economic Performance (CEP). http://cep.lse.ac.uk/pubs/download/dp1738.pdf

Krekel, C., & Zerrahn, A. (2017). Does the presence of wind turbines have negative externalities for people in their surroundings? Evidence from well-being data. *Journal of Environmental Economics and Management, 82*, 221–238.

Laffan, K. (2020). Green with satisfaction: The relationship between pro-environmental behaviours and subjective wellbeing. In D. Maddison, K. Rehdanz, & H. Welsch (Eds.), *Handbook on wellbeing, happiness and the environment* (pp. 329–348). Edward Elgar Publishing.

Lancée, S., Veenhoven, R., & Burger, M. (2017). Mood during commute in the Netherlands: What way of travel feels best for what kind of people? *Transportation Research Part a: Policy and Practice, 104*, 195–208.

Lange, F., & Dewitte, S. (2020). Positive affect and pro-environmental behavior: A preregistered experiment. *Journal of Economic Psychology, 80*, 102291.

Lerner, J., Li, Y., & Weber, E. (2012). The financial costs of sadness. *Psychological Science, 24*(1), 72–79.

Levinson, A. (2012). Valuing public goods using happiness data: The case of air quality. *Journal of Public Economics, 96*(9–10), 869–880.

Lykken, D., & Tellegen, A. (1996). Happiness is a stochastic phenomenon. *Psychological Science, 7*(3), 186–189.

Martin, L., Witte, M., Hunt, A., Richardson, M., Pahl, S., & Burt, J. (2020). Nature contact, nature connectedness and associations with health, wellbeing and pro-environmental behaviours. *Journal of Environmental Psychology, 68*, 101389.

Masson-Delmotte, V., Zhai, P., Pörtner, H. O., Roberts, D., Skea, J., Shukla, P., Pirani, A., Moufouma-Okia, W., Péan, C., Pidcock, R., Connors, S., Matthews, J., Chen, Y., Zhou, X., Gomis, M., Lonnoy, E., Maycock, T., Tignor, M., & Waterfield, T. (2019). Global warming of 1.5°C: An IPCC Special Report on the impacts of global warming of 1.5 °C above pre-industrial levels and related global greenhouse gas emission pathways, in the context of strengthening the

global response to the threat of climate change, sustainable development, and efforts to eradicate poverty (Technical report). Intergovernmental Panel on Climate Change (IPCC).
Neve, J. E. D., Clark, A., Krekel, C., Layard, R., & O'Donnell, G. (2020). Taking a wellbeing years approach to policy choice. *BMJ, 371*, m3853.
Nisbet, E., & Zelenski, J. (2011). Underestimating nearby nature: Affective forecasting errors obscure the happy path to sustainability. *Psychological Science, 22*(9), 1101–1106.
Nisbet, E., Zelenski, J., & Murphy, S. (2011). Happiness is in our nature: Exploring nature relatedness as a contributor to subjective well-being. *Journal of Happiness Studies, 12*, 303–322.
O'Donnell, G., Deaton, A., Durand, M., Halpern, D., & Layard, R. (2020). Wellbeing and policy (Technical report). Legatum Institute.
Olsson, L., Gärling, T., Ettema, D., Friman, M., & Fujii, S. (2012). Happiness and satisfaction with work commute. *Social Indicators Research, 111*, 255–263.
Onwezen, M., Antonides, G., & Bartels, J. (2013). The Norm Activation Model: An exploration of the functions of anticipated pride and guilt in pro-environmental behaviour. *Journal of Economic Psychology, 39*, 141–153.
Prati, G., Albanesi, C., & Pietrantoni, L. (2017). Social well-being and pro-environmental behavior. *Human Ecology Review, 23*(1), 123–140.
Pritchard, A., Richardson, M., Sheffield, D., & McEwan, K. (2020). The relationship between nature connectedness and eudaimonic well-being: A meta-analysis. *Journal of Happiness Studies, 21*, 1145–1167.
Pyone, J., & Isen, A. (2018). Positive affect, intertemporal choice, and levels of thinking: Increasing consumers' willingness to wait. *Journal of Marketing Research, 48*(3), 532–543.
Rees, J., Klug, S., & Bamberg, S. (2015). Guilty conscience: Motivating pro-environmental behavior by inducing negative moral emotions. *Climate Change, 130*, 439–452.
Rehdanz, K., & Maddison, D. (2008). Local environmental quality and life-satisfaction in Germany. *Ecological Economics, 64*(4), 787–797.
Riepl, K., Mussel, P., Osinsky, R., & Hewig, R. (2016). Influences of state and trait affect on behavior, feedback-related negativity, and P3b in the ultimatum game. *PLOS ONE, 11*(1), e0146358.
Schmitt, M. T., Aknin, L. B., Axsen, J., & Shwom, R. L. (2018). Unpacking the relationships between pro-environmental behavior, life satisfaction, and perceived ecological threat. *Ecological Economics, 143*, 130–140.
Schneider, C., Zaval, L., Weber, E., & Markowitz, E. (2017). The influence of anticipated pride and guilt on pro-environmental decision making. *PLOS ONE, 12*(11), e0188781. https://pubmed.ncbi.nlm.nih.gov/29190758/
Stiglitz, J., Sen, A., & Fitoussi, J. P. (2009). *Report by the commission on the measurement of economic performance and social progress* (Technical report). https://ec.europa.eu/eurostat/documents/8131721/8131772/Stiglitz-Sen-Fitoussi-Commission-report.pdf
Taufik, D., Bolderdijk, J., & Steg, L. (2015). Acting green elicits a literal warm glow. *Climate Change, 5*, 37–40.
Tiedens, L., & Linton, S. (2001). Judgment under emotional certainty and uncertainty: The effects of specific emotions on information processing. *Journal of Personality and Social Psychology, 81*(6), 973–988.
Van Praag, B., & Baarsma, B. (2005). Using happiness surveys to value intangibles: The case of airport noise. *Economic Journal, 115*(500), 224–246.
Venhoeven, L., Bolderdijk, J., & Steg, L. (2013). Explaining the paradox: How pro-environmental behaviour can both thwart and foster well-being. *Sustainability, 5*, 1372–1386.
von Möllendorff, C., & Welsch, H. (2017). Measuring renewable energy externalities: Evidence from subjective well-being data. *Land Economics, 93*(1), 109–126.
Wang, E., & Kang, N. (2018). Does life satisfaction matter for pro-environmental behavior? Empirical evidence from China General Social Survey. *Quality & Quantity, 53*, 449–469.
Welsch, H., Binder, M., & Blankenberg, A. K. (2021). Green behavior, green self-image, and subjective well-being: Separating affective and cognitive relationships. *Ecological Economics, 179*, 106854.

Welsch, H., & Khling, J. (2010). Pro-environmental behavior and rational consumer choice: Evidence from surveys of life satisfaction. *Journal of Economic Psychology, 31*(3), 405–420.

Welsch, H., & Khling, J. (2011). Are pro-environmental consumption choices utility maximizing? Evidence from subjective well-being data. *Ecological Economics, 72*, 75–87.

Welsch, H., & Khling, J. (2018). How green self image is related to subjective well-being: Pro-environmental values as a social norm. *Ecological Economics, 149*, 105–119.

Whitburn, J., Linklater, W., & Milfont, T. (2018). Exposure to urban nature and tree planting are related to pro-environmental behavior via connection to nature, the use of nature for psychological restoration, and environmental attitudes. *Environment and Behavior, 51*(7), 787–810.

Wild, K., & Woodward, A. (2019). Why are cyclists the happiest commuters? Health, pleasure and the e-bike. *Journal of Transport & Health, 14*, 100569.

Chapter 12
Conclusion of Linking Sustainability and Happiness: Theoretical and Applied Perspectives

Scott Cloutier, Sara El-Sayed, Allison Ross, and Melanie Weaver

Thank You

Hello Friend!

We hope you enjoyed the book, whether you read one chapter or the entire collection of chapters. As mentioned in the introduction, we were fortunate to have such a diverse pool of international scholars and practitioners contribute to our book. We want to take a moment to honor our contributors and all their ancestors and teachers, who's time, energy, and love supported their development and perspectives within. Please take all that you have learned and pass it forward in service to your personal and our collective sustainability and happiness. In honor of Mother Earth and all its beings, we thank you.

Love,

Scott, Sara, Ali, and Melanie

S. Cloutier (✉) · S. El-Sayed · A. Ross
Arizona State University, Phoenix, AZ, USA
e-mail: Scott.Cloutier@asu.edu

S. El-Sayed
e-mail: Sara.elsayed@asu.edu

A. Ross
e-mail: Allison.Poulos@asu.edu

M. Weaver
School, Ann Arbor, MI, USA
e-mail: Melanie.Weaver@asu.edu

S. Cloutier et al. (eds.), *Linking Sustainability and Happiness*,
Community Quality-of-Life and Well-Being,
https://doi.org/10.1007/978-3-030-89559-4_12

12.1 An Honoring

As we conclude our book, we want to reiterate that our work is in no means beyond prior works, even with seemingly new findings. We honor the works prior and seek to add to the conversation both within and outside of academic spaces. We ultimately see the book as a contribution to our responsibility to be aligned with our planet and an opportunity to (re)align with our ancestors to find a shared path toward sustainability and happiness.

12.2 Sustainability and Happiness Defined

As you read, there are many perspectives on and definitions of sustainability and happiness, each of which carries its own importance. In general, the editors think of sustainability and happiness as both nouns and verbs. Sustainability noun, also a destination, can be thought of as a place where regenerative living is the norm, and we relate to all beings in reciprocity, respect, and a deep awareness of our reliance on and responsibility to natural systems. Thus, sustainability verb is the process of enacting such a vision to reach a sustained state of harmony with natural systems. The same can be said for happiness which, as a noun and a destination, can be thought of as a sense of wellbeing grounded in a life of joy and contentment. Happiness, as a verb and a process, is the practice of moving toward happiness noun, learning the steps and actions needed to find a joyful life.

12.3 Integrating Sustainability and Happiness Perspectives

Our contributors offered several sustainability and happiness perspectives, from both theoretical and applied framings. In total, we had 10 chapters, four of which are theoretical and six applied. Our theoretical chapters made the case for alternative theories that can simultaneously enhance sustainability and happiness. Theoretical perspectives included (Chap. 1) promoting community productivity by Maria Spiliotopoulou and Mark Roseland, (Chap. 2) considering means of promoting life by Jacob Bethem, (Chap. 3) including internal dimensions of happiness in sustainable development processes by Sauwalak Kittiprapas, and (Chap. 4) promoting Sarvodaya (Welfare for All) by Jorge Guardiola.

Our applied chapters consisted of six studies investigating the links between sustainability and happiness. Applied perspectives included (Chap. 5) testing a double dividend relationship between happiness and pro-environmental behaviors by Nazaret Ibáñez-Rueda and Jorge Guardiola, (Chap. 6) investigating how the

SDGs and various environmental indicators interact with the happiness and well-being measures by Leire Iriarte from El Buen Vivir, (Chap. 7) identifying interesting factors affecting women's decisions to participate in the labor market, or not, by Shoirakhon Nurdinova, (Chap. 8) investigating happiness and sustainable development (environmental presentation) relationships in Thailand by Sauwalak Kittiprapas, (Chap. 9) exploring the relationships between happiness indices and sustainable mobility measures in the world's happiest cities by Alshimaa Aboelmakarem Farag, and (Chap. 10) linking happiness and pro environmental behavior by Christian Krekel and Alberto Prati.

In sum, our chapters blend numerous perspectives on both sustainability and happiness. Some found that subjective aspects of wellbeing (i.e. life satisfaction) have been correlated with sustainability metrics (i.e. pro environmental behavior, sustainable development, environmental preservation), though deeper conceptions of happiness like eudamonia and internal aspects of happiness (i.e. inter-personal life satisfaction, inner peace) have not been linked to sustainability. The Sustainable Development Goals (SDGs) were analyzed and critiqued with the intention of explicitly promoting happiness. Other contributions make a case for reconsidering the deeper purpose and intentions behind sustainability and happiness. Approaches include trying to live life in a better way grounded in fundamental concepts of good living like acting with respect for others' vitality, acting in love, acting peacefully by honoring Sarvodaya, and through better relations with the natural world by supporting the flourishing of non-humans. Studies also found that those cities that support aspects of sustainable living (i.e. sustainable transportation infrastructure) have happier residents, and that improving the productivity of community assets can promote community happiness and wellbeing.

What we, the editors, have found is that most theories and applied findings recognize our reliance on natural systems and the happiness that comes beyond material consumption and short-term dopamine releases from unsustainable behavior. More, the chapters recognize a good deal of work yet to be done to realize a sustainable state and the processes needed to get there. Cloutier (editor 1) suggests that happiness can be the means of actualizing sustainability. In other words, the process of moving toward sustainability and experiencing sustainable states can feel good and be rewarded by our body's evolutionary machinery. Up to this point, much of the rhetoric around sustainability is problem-based, identifying the many reasons we have made mistakes, and why we should be fearful, shameful and guilty. A shift is needed and is possible by more closely assessing the relationships between sustainability and happiness. In this way, sustainability noun can be aligned with happiness noun and both sustainability and happiness verbs can be the complementary processes of moving toward a sustainable and happy future.

12.4 Closing

We thank you for reading our book and hope you will take away whatever is needed to contribute to a brighter future for our planet and its beings. We honor you and all your ancestors and teachers for what they have offered. We see this book as the continuation of conversations and investigations into realizing a sustainable and happy future and seek to form a collaborative international working group focused on sustainability and happiness relationships. If interested, reach out to our team and say hello. We foresee this as some of the most important work of our time and welcome all minds, hearts, and souls to the table to contribute.

With Love,

The Editors

The manufacturer's authorised representative in the EU is Springer Nature Customer Service Centre GmbH, Europaplatz 3, 69115 Heidelberg, Germany. If you have any concerns regarding our products, please contact ProductSafety@springernature.com

Printed and bound by CPI Group (UK) Ltd, Croydon, CR0 4YY
30/09/2026
02225508-0002